Keeping Pigs

A PRACTICAL GUIDE FOR SMALLHOLDERS

By Claire Scott, James Adams
and Peter Siviter

With special contribution from Sue and Stephen Dudley

First published in 2023 by
The Crowood Press Ltd
Ramsbury, Marlborough
Wiltshire SN8 2HR

enquiries@crowood.com

www.crowood.com

British Library Cataloguing-in-Publication Data
A catalogue record for this book is available from the British Library.

ISBN 978 0 7198 4257 3

Typeset by Envisage IT

Cover design by Blue Sunflower Creative

Printed and bound in India by Thomson Press Limited

CONTENTS

	Introduction	6
1	Pig Choices	8
2	Legislation and Regulation around UK Pig Keeping	27
3	Biosecurity	30
4	Pig Housing and Environment	36
5	Feeding	50
6	Handling	62
7	Transporting Pigs	73
8	Making the Most of Your Vet	79
9	Diseases of Pigs	93
10	Medicines	125
11	Vaccinations	138
12	Parasite Control	142
13	Breeding	147
14	Pig to Pork	181
15	Common Procedures	189
16	Pig Death and Euthanasia	202

Conclusion	208
Glossary	209
References and Resources	211
About the Authors	218
Index	220

INTRODUCTION

The British pig can be characterised as a determined protagonist in a constantly evolving story. After World War II, a spate of notifiable disease and animal farming crises arose. 'Mad cow disease', 'Foot and Mouth' and salmonellosis led to distrust in the commercial meat-producing market. Alongside this, concerns of animal welfare on farms came to the fore. This has resulted in three movements that are of intense relevance to this book. They are as follows:

Pigs deserve better: Rearing and growing your own food is, to many, the best way to ensure confidence in its production. Barbara and Tom have inspired the masses to create their very own 'Good Life'. Pigs, and buyers of their produce, should expect dignity and individual care throughout life and in death. In addition, such holdings are crucial for the preservation of rare breeds. This might sound a noble and romantic endeavour, but don't forget that it can only be achieved with much hard work.

Tightening of red tape: In response to mistrust in commercially reared meat, the agricultural sector has strived to reassure consumers with changing legislation, introduction of farm assurance schemes, and strict supermarket contracts.

Pigs as equals: A growing group has moved towards a goal of equality amongst all living creatures, resulting in pigs occupying spaces in animal sanctuaries, gardens, stables, garages, conservatories and living rooms. They can be companions, best friends, surrogate children, or simply pasture ornaments.

Sue and Stephen's pigs are raised for meat, but are kept with intense love and dedication.

So why is it important that our readers acknowledge these three dimensions? Firstly because it got us here, in the context of this book, a book that hopes to be read by, and to cater for, both the pet pig owner and the small-scale pig producer. This is no mean feat we are sure you will agree, but we hope

These pet pigs have no 'job' other than to provide amusement and joy to those around them.

that readers will identify with our common goal: to encourage the keeping of the happiest and healthiest pigs possible, no matter what their intended function or lifespan might be.

Secondly, because it is important for you to know what you are up against, given that most smallholders aim to provide a better life for a pig than the life they would receive on a commercial pig farm. Therefore we will make reference to commercial farming, because we have learnt so much about pigs by farming them for so many years.

Throughout, we have strived to provide the most current and relevant information for smallholders keeping pigs in the UK. Due to the quickly moving and progressive nature of the pig sector, much of this has changed even during the two-and-a-half year writing process for this book. Therefore, readers must ensure that they are using the most up-to-date practices and following current legislation for their area. This will generally be achieved by consulting the extensive sources that we provide as reference material. Whilst we will try to update the book where possible, we can't be held liable for this changing climate.

In the following chapters we will focus on preventing pigs from suffering from disease, which we regard as the best way to ensure happy and healthy lives. This encompasses ensuring that pigs have suitable nutrition and appropriate housing, as well as using vaccines and medicines appropriately.

By preventing health problems, first and foremost we protect our pigs from unnecessary pain and suffering. Secondly, preventing disease saves money, for example by avoiding the costs of emergency veterinary visits and unnecessary medicines, or by allowing pigs to achieve their growth and reproductive potential. Finally, treating disease is often attempted using an antibiotic. Preventing disease is therefore an excellent way to reduce the risk of the development of antibiotic resistance, which is a global health concern.

In this way we aim to equip you with some of the knowledge and practices that you will need to keep pigs. Mistakes will be made along the way, but learning from these will ensure a higher standard of welfare on your holding, and we urge you to keep striving for the best that you can possibly achieve.

These piglets are provided with everything they could need.

CHAPTER 1

PIG CHOICES

SHOULD I KEEP PIGS?

Before sourcing pigs, you must consider whether you can provide an appropriate home for them, by evaluating whether you can meet their needs. Keepers of pigs are required by law to have a copy of, and be familiar with, the 'Code of Practice for the Welfare of Pigs'.[1] Much of this book has been informed by the Code of Practice (which is how we will refer to it from now), but we aim to translate it into what is pertinent for smallholders and provide the tools to meet it.

The Code of Practice is built round a framework called 'the Five Freedoms', devised by the Farm Animal Welfare Council - FAWC - in 2007,[2] which describes the needs of farmed animals. All pigs, even pets, are covered by farm animal legislation in the UK and are therefore protected by this framework. They are:

1. Freedom from hunger and thirst by ready access to fresh water and a diet to maintain full health and vigour.
2. Freedom from discomfort by providing an appropriate environment including shelter and a comfortable resting area.
3. Freedom from pain, injury or disease by prevention or rapid diagnosis and treatment.
4. Freedom to express normal behaviour by providing sufficient space, proper facilities and company of the animals' own kind.
5. Freedom from fear and distress by ensuring conditions and treatment to avoid mental suffering.

By law, all keepers must have a copy of, and be familiar with, Defra's Code of Practice for the welfare of pigs.[1]

Seemingly simple on the surface, these can be difficult to achieve simultaneously. To give an example: on indoor commercial pig farms, farrowing sows are often housed in 'crates' with their piglets for up to four weeks after the piglets have been born. This allows sow and piglets to be closely monitored, and the keeper can safely intervene when required. The 'crate' is designed so that piglets are not squashed and killed when the sow lies down. Rooms are temperature controlled. A heat mat and lamp is provided in a creep area for the piglets to keep them warm, while the sow lies on metal to allow her heat to exude from her. She is fed precisely and her intake monitored closely, as are her piglets. Slatted floors allow faeces to drop away from the pigs, meaning that disease from these faeces is much less likely to pass into the mouth of another pig.

However, these pigs have little opportunity to display natural behaviours. Sows cannot turn round in their crates, and they cannot interact with each other. Slats mean that sows can only be given limited nesting material, as too much straw would block up the slats. Therefore, despite so many measures intended to look after the sow and piglets, her freedoms are compromised for that period.

Conversely, a farrowing sow kept on an outdoor commercial pig farm (about 40 per cent of UK pig production[3]) will generally be housed in a tin or plastic pig arc, which in the summer can be sweltering and in the winter will be fairly chilly - however, body heat does a good job at keeping her warm. Monitoring is limited to ensuring that every sow has come out to eat each morning, at which point piglets are briefly checked. In the summer the ground is dusty and rock hard, while in winter sows

This more commercial outdoor pig farm looks great when the sun is shining, but can be much more difficult to manage in adverse weather.

may be up to their nipples in mud just to leave their arc to eat, drink and defecate. Despite these challenges, these sows and their piglets live with the freedom to act like pigs: to root in the dirt, to wallow, and to frolic in the sun.

In this way, it is important to note that where improvements are achieved in one of the Five Freedoms, there may be a cost to another. It is up to keepers to balance these factors within legal and best practice frameworks.

Is one system better than the other? That is likely to be highly personal. We believe that, once the basics have been achieved, it is good stock(hu)manship that is key to well-kept pigs, rather than the system in which they are kept. To achieve this, pig keepers should embody FAWC's essentials of stockmanship 365 days of the year (pigs do not care that it is a Sunday morning). These are:[2]

Knowledge of animal husbandry: Sound knowledge of the biology and husbandry of farm animals, including how their needs may be best provided for in all circumstances.

Skills in animal husbandry: Demonstrable skills in observation, handling, care and treatment of animals, and problem detection and resolution.

Personal qualities: Affinity and empathy with animals, dedication and patience.

Unfortunately good stockmanship isn't always this idyllic!

Only if you feel that you are able to meet these qualities and fulfil pig needs in a considered and caring way, should you proceed with the idea of keeping pigs.

Before sourcing pigs, it is really important that readers at least understand the basics of pig care to ensure that a suitable home is achieved.

THE SMALL-SCALE PRODUCER

We recommend starting out with two or three recently weaned pigs to rear and send for slaughter. There is so much to learn concerning pig rearing before breeding should be embarked upon. When choosing your weaners, the following questions should be considered:

How many? Pigs are social creatures and should not be kept alone, so always buy more than one.

What sex? Initially, buy male pigs as you are much less likely to become attached to them and to keep them for breeding. The decision to send your first pigs to the abattoir can be tough - however, adolescent male boars will make that decision as easy as possible. To avoid pregnancies, buy pigs of the same sex. Pigs breed very early and may take you by surprise - and pigs don't have the same social boundaries as humans, so will breed with their siblings without worry.

Rearing three weaners is preferable to two, in case something disastrous happens to one.

BOAR TAINT

Well-meaning pig keepers may warn against male pigs due to 'boar taint'. Boar taint is the smell or taste of male pig meat that can occur from the production of male sex hormones. It is very unusual in rare breed pigs, probably due to their slow maturation rate. Rare breeders we meet have never noticed it, even those who have killed boars that have actually 'worked' (served sows). Therefore it is not a good reason to request female weaners or to castrate male weaners.

Even a working male such as this boar at Millhouse Nursery is unlikely to exhibit boar taint.

When? Start by rearing pigs over the summer, but try a winter batch before breeding any pigs to see how your system copes with the challenges that adverse weather and short days bring.

Where From?

Rearing weaner pigs for meat is critical to the survival of rare breeds. Only the best pigs from each litter should be used for breeding, and therefore the others are destined for a dinner table. The phrase 'the only way to keep rare pig breeds is to eat them' really is true. Beyond that, we believe that choice of breeder is more important than choice of breed. You will have time to try different breeds before you think about breeding any pigs yourself, and therefore we would most like you to find the perfect breeder, in preference to the perfect breed.

When deciding on a breeder look for someone who:

- birth notifies their piglets: without this, pigs may well not be pedigree and you are not acting to preserve a breed.
- is local to you: not only will logistics such as your initial movement be easier, but you can benefit from their network and utilise relationships that they have built up, such as a vet or an abattoir.
- provides mentorship and support: good breeders want to know that their pigs are going to an excellent home.

Several breeders also run pig courses where you can get to know their pigs, learn about pig management, and ask lots of questions. Sue and Stephen's course promises that you will be Large Black enthusiasts by the end of the day!

Alternatively, contact a few breeders and arrange to see their holding. To make sure that you aren't spreading diseases, leave at least three days between visits, and wear clean clothes, and thoroughly clean wellies.

If you are not happy that their pigs' needs are met, walk away. You don't want your first pig-keeping experience to end in tears. If you want to rescue a pig, there are many farm animal sanctuaries happy to assist you.

QUESTIONS TO ASK THE BREEDER

The following questions are useful to discuss with breeders:

Timing

- When are they expecting litters? They should have a good idea if running a tight ship.
- At what age do they wean?
- At what age do they sell weaners?
- At what age do they sell pigs for market?

This will give you an idea of how long you will have to keep the pigs before they are finished.

Networks

- Do they have a good relationship with a vet that you could also use? Be wary of keepers who don't advocate cultivating a good working relationship with a vet.
- What abattoir do they use?
- How do they get there?
- Do they have a trailer they could lend you for the day of marketing?

Husbandry

- What do they feed the pigs?
- How do they enclose their pigs?
- Will pigs arrive electric-fence trained?
- What is their health routine, especially in terms of vaccines, worming and husbandry procedures?
- Are there any diseases to be aware of?

KEEPING PIGS AS PETS

The decision to keep pigs as pets is very different from that of a producer, not because they are different animals, but because we place different requirements on them and often place them in environments that are not designed for a pig. Therefore very specific attention must be paid to meeting their specific species needs so that unintended issues do not arise over the many, many years that they are likely to live for.

Pigs have a need to express natural behaviour, such as rooting and investigating their space as well as roughly playing with pen mates. Treating pigs inappropriately, or not giving them a suitable outlet for these behaviours, is likely to lead to behavioural 'problems' such as aggression or destruction. Punishing pigs that express these behaviours will only make these worse.

It is also likely that others will not see your pig as they would a dog. A pig's noisy and destructive natural behaviours may lead to issues around neighbours' complaints, lack of holiday care, unsympathetic landlords, and life changes meaning that a pig is no longer suitable for you.

For these reasons, rescue organisations are currently filled with pigs purchased as pets that were no longer suitable for their first home. Many will be 'micropigs' that outgrew their name. Many will be accidental families from brother-sister matings. Others will be in dire straits after being given completely inappropriate care in their first home. Therefore it is imperative to do the appropriate research before bringing home pet pigs. A large part of this will be considering how you will allow your pet pigs to be pigs.

If you are considering pet pigs, please reach out to a rescue organisation or farm

Even pet pigs are best kept with other pigs in grassy paddocks where they can express natural behaviours.

animal sanctuary near you. By purchasing from a rescue organisation you are likely to acquire an older pig that is more predictable behaviourally. For example, if you would like pigs that are less likely to destroy ground, a sanctuary can show you pigs that do not do this.

Adopting a fully grown pig will also be preferable, to avoid any 'micropig' mishaps where a pig grows to a much larger size than expected. A 'minipig' or 'micropig' is not a breed, but a mixture of small pig breeds where the smallest pigs from each litter were bred together over generations (often also resulting in genetic problems from inbreeding). This means that 'large pig' genes can still be in the mix and can crop up sporadically. Furthermore, unscrupulous breeders will lie about the age of pigs or hugely restrict feeding of their own pigs so that their growth is stunted. In different hands these pigs can grow to be much bigger!

There are some fantastic farm animal sanctuaries where those who run them will mentor and guide you to ensure that you will make an appropriate home for their pigs - but be sure always to do your own research, and check advice with your vet.

ANIMAL SANCTUARY CHECKS

Before giving you a pig, a farm animal sanctuary should do the following:

- Check your premises, and not rehome to a holding where pigs will be kept exclusively indoors.
- Demand details such as a CPH number and other legislative adherence.
- Not rehome lone pigs unless there is a very specific reason in a particular pig.
- Be able to discuss pig-specific husbandry needs at length.

Wherever you do source pet pigs, it is absolutely imperative that pigs are not housed alone. They are highly social creatures that need other pig company. Housing pigs alone will eventually cause behavioural and aggression problems. Pet male pigs must be castrated to avoid unwanted pregnancies and to reduce boarish behaviour, such as aggression or mounting, later in life.

BREED CHOICES

As discussed, we believe that your breeder choice and future mentor is far more important than the choice of breed. However, we also think that preserving rare breeds is an admirable part of small-scale pig keeping, and therefore we would like to devote some time to discussing some of the breeds available to you. So whilst keeping weaners, before you consider branching into the world of breeding, keep a few different breeds from different breeders (not at the same time - you shouldn't mix weaners from different sources for health reasons) and see which you prefer and what suits your system.

Informing Your Choice of Breed

Size and Character

Big pigs are powerful, and some breeds are boisterous. Your own situation will inform what is appropriate in this respect, and whether you want a smaller breed such as a Middle White or Berkshire, or a large pig such as a Gloucester Old Spot or Large White. Obviously, any pig breed can be kept as a pet given appropriate conditions, but a very large pig is far less practical. The Kunekune, for example, is a classic pet pig breed.

Rooting

Do you want pigs to clear a patch of land, or would you like the least destruction possible? Bear in mind that any pig can root, but in general, the shorter its snout, the less power and destruction it will bring, looking to the Kunekune and Middle White as examples.

Rarity

The vast majority of commercial pig producers breed mixed-breed pigs and rarely have pure bloodlines on farm. Therefore our only credible chance of maintaining these pure breeds is through our network of small-scale pig producers. Details of these rare breeds are found from the Rare Breeds Survival Trust (RBST).[4] Do remember that rare things are often rare for a reason, and you may need to accept some poor litter numbers or conception rates from the rarest options.

Rooting and investigating the ground around them is a natural behaviour for pigs.

If this is your intent, please only buy stock from breeders who register their pedigrees. If they are not registered, they are worthless to this intent. There are different rules for different classes of BPA members, but in essence the following rules apply:

- Pigs that could be used for breeding should be bought from a pedigree breeder. Each pig should be birth notified and identified by an individual ear number.
- Weaners bought for fattening should be birth notified but need not be individually identified. This is to ensure that the parents are pedigree and that the breed is therefore being preserved. Buying cross-breeds as fattening weaners will still act to preserve rare breeds as long as both parents are pedigree. However, these pigs cannot be herdbook-registered breeding stock and must not be bred from in order to maintain the breed.
- When weaners are bought, they must be transferred to you on the BPA Grassroots database, which will enable you to sell the meat with 'Pedigree Pork' certificates. To sell pork as 'Pedigree Pork' you also need to have joined the BPA's Pedigree Pork Scheme.

Meat Characteristics

We have yet to meet a breeder who doesn't think that their pig meat is the best, but some will profess to extra qualities such as fat marbling.

SOME RARE UK PIG BREEDS

Once you have narrowed your choice into some kind of specification, you may find it helpful to build up a certain knowledge on each breed. The details of many breeds are found through the British Pig Association (BPA), although British Kunekunes and British Lops have their own separate societies. We have also provided a brief snapshot of each breed from some of our most esteemed keepers and champions of each breed. We thank them for their amazing contributions.

The Berkshire

Thought to have originated around the Thames Valley in the 1790s, the Berkshire was a large, tawny red pig, spotted with black, with a long body, ears that were lopped, short in the leg, and larger and coarser than today's Berkshire. Asian blood was introduced in the 1800s, making the body shorter and deeper, and the ears much smaller and more in line with what we see today. Modern Berkshires must have six white points, namely the face, all four feet and the tail. They have fine, dished faces of medium length and fairly large pricked ears; the head should not be coarse.

They are kind, docile pigs that produce good litters and milk well, and are easy to manage. One of the smaller of the traditional breeds, they are quick to finish to pork weight, killing out at between 60 and 70kg. They are good food converters, and do well whether living in or out. Like any pig they will root, and so enough space to rotate their rooting ground is needed. The delicious meat is very sought after by butchers and restaurants. Popular with the Japanese, Berkshires command a good price as Kurobuta pork. Pigs producing this pork must

A Berkshire sow.

have a pedigree from an established herd in either England or Japan. The fat is marbled throughout each cut, allowing the meat to cook evenly and giving a much juicier result; it also has a sweeter taste.

Sharon Barnfield, Kilcot Pedigree Pigs

The British Landrace

The British Landrace is a white breed with short, lopped ears. The breed first came to the UK from Sweden in 1949. Other imports followed from 1953 onwards, and the British Landrace Society was created. New bloodlines were then imported in the 1980s from Norway. The breed expanded rapidly, and is one of the UK's most popular breeds, however it has suffered with low numbers of pedigree registrations. With only twenty-six registered pedigree breeders, sow numbers have now dropped to 121, with only 152 registered from the last breed survey in 2021; they are now on the RBST priority list.

The breed performs well in both indoor and outdoor management systems. Sows are able to rear large litters, and make fantastic mothers. The Landrace has a good temperament, but sows can be very protective of their piglets, so are best suited to more experienced pig keepers.

One of the greatest strengths of the British Landrace is its undisputed ability to improve other breeds when used to produce hybrid gilts. Over 90 per cent of hybrid gilts in Western Europe and North America contain a Landrace bloodline to help with meat production. Piglets have a good daily weight gain and a high lean-meat content. They have a superbly fleshed carcass, and are ideal for bacon and pork production.

Grace Bretherton, Junior Pig Club Youth Ambassador

The British Lop

As its original name suggests, this pig is long and white, with lop ears reaching to the tip of its nose. The British Lop Pig Society was formed (as the Long White Lop-eared Pig Society) in 1920, and has remained independent ever since, making it the cheapest route to pedigree registration, with both membership and registration fees very low.

The breed is very docile and easy to manage; indeed, a single strand of electric fence wire some 30cm above the ground will confine these pigs. Several years ago the writer's son showed a Lop sow at the age of three! The breed is noted for its large litters of fourteen to sixteen piglets, but it is clumsy, and two or three might be lost when the sow lies down.

Piglets should reach pork weight in about sixteen weeks, when there will be under 10mm of fat cover. Bacon weight is achieved

A fine British Landrace sow.

A prize-winning British Lop boar.

in about twenty-six weeks, when there will be around 25mm of fat. Most importantly, the skin and hair are white so there is no discrimination by the butcher. Conformation is good, with the length and ham size that the butchers prefer. The meat is succulent with intramuscular fat, and in several taste tests has come out top.

Frank J. Miller, former Chairman and Secretary of the British Lop Pig Society

The British Saddleback

The Saddleback is a semi lop-eared pig with ears facing forwards; it is black, with a band of white hair running from one foot over the shoulders and back to the other foot. There can be white on the nose, the tip of the tail and the back feet. The white hind feet are a throwback to the Essex Saddleback, dating from the time before the Essex and Wessex Saddlebacks were amalgamated in 1967.

The Saddleback sow must have twelve evenly spaced teats, and often has large litters - our first sow, 'Stripe', had nineteen in her first litter and raised them all! She gave us a great introduction to showing as it was so easy to do anything with her - and indeed, this is a very strong characteristic of the breed.

Saddlebacks are fantastic mothers, generally being very careful, and they happily tolerate having humans around the piglets. We farrow our Saddlebacks both indoors and out successfully, with and without heat lamps; the average litter is ten to twelve piglets, but is often more. They love grazing; ours have paddocks of about an acre for three, and they also have a wallow to play in.

The Saddleback produces wonderful meat, be it joints, hog roasts, sausages, hams or bacon; it is truly versatile. The weaners reach pork weight in approximately sixteen weeks.

In my opinion the Saddleback is the best pig for those new to the world of pigs.

Sharon Groves, committee member British Saddleback Breeders' Club, BPA British Saddleback breed representative

The Duroc

A large-framed auburn pig, the Duroc couldn't be better suited to the smallholder. Originally from America, Durocs were derived from the early 'Red Hogs'. The breed was unsuccessfully introduced to the UK in the early 1970s, then a decade later was successfully re-imported. It is now popular worldwide in the commercial sector.

Attentive and protective mothers, litter sizes range from ten to fourteen; however, for safety, minimal piglet handling is advised - but when they are not suckling, the sows are amongst the most docile. Their thick winter coat ensures they are able to cope with cold, wet winters.

Commonly used as a terminal sire, the Duroc can also be used this way on the

Champion Saddleback boar 'Watchingwell Dominator 130', otherwise known as 'Dylan'.

Duroc sow 'Portbredy Lena 1337'.

smallholding. Thus crossing a Duroc boar with a traditional sow that is cheaper to maintain with less feed required will improve the meat quality of the offspring, with better marbling, flavour and growth rates.

The Duroc grows fast, with good muscle quality, and fat that is marbled throughout the muscle rather than in a layer under the skin. The Duroc can achieve a deadweight of 60kg by eighteen weeks of age, with a subcutaneous fat measurement of less than a quarter of an inch. Durocs need a high-protein, high-energy compound feed. Fruit and vegetables simply limit the feed intake and inhibit potential.

Dr Oliver Giles BVMSci, MRCVS, The Tedfold Herd of Pedigree Pigs

The Gloucester Old Spot

The Gloucester Old Spot is a large pink pig with black spots: these should cover less than 50 per cent of the body. They have a dished face, and lop ears that reach almost to the end of the nose. Spotted pigs were first recorded in 1913. Called Old Spots, they originated around the Berkeley Vale area and were kept in cider orchards and on dairy farms where they were fed on the waste from these enterprises. Folklore has it that the spots are from the apples falling and leaving bruises on the backs of the pigs.

Gloucester Old Spots are good milky mothers, producing litters of eight or more, and are docile around farrowing, which makes them a good beginner's pig. On top of this, they are attractive pigs, making them sought after by smallholders. All pigs will root and Gloucester Old Spots are no exception, but a good-sized enclosure or two will allow you to rotate them to let the ground recover.

A Gloucester Old Spot sow.

They are a good dual-purpose pig, ideal for producing both pork and bacon, and they are easy to manage, happily living either in or out. They convert food well, and grow to either pork or bacon weight without an excessive amount of expensive concentrate food. The meat is much in demand, making them a good breed on the selling front.

Sharon Barnfield, Kilcot Pedigree Pigs

The Kunekune

Originating from New Zealand, Kunekunes were brought to the UK in the 1990s and quickly became popular. However, there are only seven female bloodlines and four male; one sow line has already been lost, and numbers of some of the other lines are diminishing, making the breed vulnerable.

There is much variation in colour and coat: they can be bald, smooth or hairy, and sometimes curly! Colours range from light cream through to black, including ginger, brown, 'gold tip' and with patches of colour.

A Kunekune sow.

Kunekunes are an ideal starter pig as they grow very slowly, making them manageable for longer. They are also docile and intelligent, and can manage on less ground than larger pigs because they are less destructive. They are good, milky mothers, farrowing with ease and producing litters of three to twelve piglets. Regarded by many as a pet pig, they are easy to train to basic commands. Their gentle nature and varied colouring make them ideal for visitor attractions, and although good mothers, they are not too protective, making it easy to interact with young piglets.

A slow grower inclined to fat, they are cheaper to rear for the table, but are not ready to slaughter until twelve to eighteen months of age at 50–70kg, providing small but delicious joints of dark meat. They also make perfect sausage pigs.

Wendy Scudamore, Former Chairman of The British Kunekune Pig Society

The Large Black

The Large Black is a large, black, lop-eared pig whose origins lie with the Old English Hog. The black hair makes them resistant to both sunburn and cold, and pre-war they were exported worldwide. Numbers plummeted post-war and became critical. The active and welcoming Large Black Pig Breeders' Club was founded in 1996 and gives advice to anyone interested in the breed. Today, Large Blacks are rated 'priority' by the RBST, but there are more breeders than might be expected as herd sizes tend to be small.

'Cerridwen', a lovely Large Black sow from the particularly rare 'Queen' line.

As favourite characteristics, friendliness and flavour both rate highly. Large Blacks have such personalities - you will be constantly laughing at their antics. On the other hand, breeding animals can attain significant size - our boar Bran is around 400kg, so simply scratching can cause damage.

They are not great rooters, and unless confined to areas that are too small, are unlikely to escape, so stock fencing is generally adequate. Quick learners, they can easily be trained to electric fences, and they are friendly, intelligent and curious, enjoying interaction with humans. They are very easy to handle despite their large size. The breed produces litters of eight to ten, and they are excellent, milky mothers.

Slow growing, they can become fat if overfed. With care, they give a dark, succulent meat, marbled with delicious fat, making the breed excellent for charcuterie. Porkers slaughter well at seven months, producing a 65–75kg carcass. Although the hair is black, the skin, and therefore the crackling, is white.

Stephen and Sue Dudley, Black Orchard Large Blacks

The Large White

The Large White is a white pig with pricked ears, a dished face and a straight back. The breed's origins are in Yorkshire, and in some countries it is known as the Yorkshire pig. Farmers started selecting white pigs with prick ears in the nineteenth century, and in 1884 the first herd book was published. From then onwards the breed increased rapidly in popularity, helped in no small part by the demand for the breed all over the world. In fact, the Large White pig

A young Large White boar from the 'Viking' line.

A blonde Mangalitza sow from Otterburn Mangalitza.

can be found in just about every country in the world where there is a pig industry. In the UK the breed is the foundation of most breeding company programmes. However, pure breed numbers have dropped dramatically in the last twenty years to stand at just 289 sows and forty-five breeders, with some female lines being perilously close to extinction.

The Large White is a good mother and will produce litters of ten or more piglets for many years in both indoor and outdoor systems. Kept outdoors, a wallow is essential in the summer months to prevent sunburn. These pigs are easy to handle and have a good temperament.

The Large White is the ultimate meat-producing pig, being ideal for both pork and bacon production. Even if kept to heavier weights they do not lay down excessive amounts of fat, and produce a carcass that grades well at all weights. They grow well, producing an ideal carcass for the modern butcher, with just the right amount of fat covering on the loin. Fed correctly they will reach 90kg liveweight in 150 days.

Nick Kiddy, Solitaire Farm, BPA Large White breed representative

The Mangalitza

An extremely hairy breed, the Mangalitza has three colours, red, blonde and swallow-belly, and when these are bred together huge variations result. Created in the 1830s, each colour produces a different type of meat. Its fat was so prized that the Mangalitza was traded on the Vienna Stock Exchange, but by the 1980s the breed was near extinction. Thankfully it was saved, but it is still rare in the UK, with only eighty-seven sows and twenty-nine boars in 2019. Mangalitzas have black trotters, snouts, teats and tail tassels, and the ears can be pricked or floppy. The piglets are born with stripes, but these soon fade.

Whilst they are perfect for smallholders, being placid and friendly, they can be challenging escapologists. They don't root, they plough and quarry, and will dig under fences. The sows can be protective, so the best thing is to leave them alone – they will introduce you to their piglets when they are ready. Litters average six, but can be as large as ten.

Slow growing, they are not ready to process until they are eighteen months old, when they yield around 120kg. They improve with age, and we have processed pigs as big as 350kg and as old as eight years. The meat is unique, deep red and much nearer beef in taste. It can be heavily marbled, and the back fat is high in monounsaturates, with the ratio of fat to meat as high as 60/40.

They are a charcutier's dream, and an amazing all-rounder for joints, sausages and bacon. The fat is perfect to add moisture to other meats, and can even be used in pastries and cakes, like butter.

Lisa Hodgson, Otterburn Mangalitza, BPA Mangalitza breed representative

The Middle White

The Middle White pig is an early-maturing pork pig, and is one of the most recognisable pigs, with its moderately short face, broad dished snout, and large pricked ears. The breed was first recognised in 1852 at a show in Yorkshire, where the judge decided that some pigs were too small for the Large White classes and too large for the Small White classes, so the 'Middle' White class was created. In the 1930s it was the most populous breed in the country, but it is now amongst the rarest British breeds. It is still farmed in Japan, where the quality of its pork is highly prized, and there has even been a shrine erected to honour the excellence of the breed. The pork is popular with chefs in the UK, and it is also used to supply the suckling pig trade.

It is a friendly pig that is easy to manage, and is therefore ideal for novices and smallholders. Sows make great mothers, milking well and raising litters averaging eight to ten in size. They are hardy, and can live outdoors all year round with no issues and don't take up a huge amount of space, but like all other pigs, they will root, and so space to rotate them on to fresh pasture and allow restoration of used ground is of benefit.

Youngsters tend to be weaned at eight weeks of age and will achieve pork weight in six months, allowing for variables, on a standard sow and weaner ration. It is also possible to take them on for a bit longer to achieve bacon weight, with careful feeding to prevent excess fat.

Cathy Baker, Rockwood Rare Breeds

The Oxford Sandy and Black

Historically referred to as the 'plum pudding pig', the Oxford Sandy and Black is one of the UK's oldest native breeds. After becoming almost extinct by the 1970s, today there are thirteen sow lines and four boar lines remaining, with approximately 420 breeding sows and 120 boars (2021/2022).

These pigs have a pale sandy to rust-coloured coat with patches of black. They have lopped or semi-lopped ears, a medium-length nose, white socks, a blaze and a tassel, and solid conformation with a good shoulder and hams, and a proud, strong, straight back.

They make excellent mothers, farrowing litters with an average of eight to twelve piglets. They are easy to manage, move and

A Middle White gilt.

An Oxford Sandy and Black boar from the Broadways herd.

handle due to their docile temperament, so are an excellent choice for a smallholder or new pig keeper. They are slow growing, and medium to large in size, and are very hardy to weather conditions. They like to live outdoors with the freedom to root and roam. An Oxford Sandy and Black will typically finish between six and eight months at 65 to 85kg.

Tania Whittick, Broadways Rare Breed Porkers

The Tamworth

Tamworths originate from Tamworth in the West Midlands. The latest breed survey (2021) showed that there are only 239 registered Tamworth sows, only 300 total registered pigs, with only seventy-nine breeders. It is the only whole-coloured ginger pig, with pricked ears and a lively curiosity.

The Tamworth is a good smallholder's pig. They make very good mothers, are very milky, and generally just get on with their task. Litters are usually a manageable size (our average is eight reared per litter), and they do it without needing expensive creep feed. With their long noses, they are good at rooting, and clear scrubland very well.

Tamworths are equally delicious for pork and bacon. Over the last twenty years or so they have improved considerably in both conformation and temperament, an achievement due to a small group of dedicated breeders. Pigs reach pork weight (50 to 70kg) between six to eight months, or bacon weight (80 to 100kg) between eight to twelve months.

Bill and Shirley Howes, founding members of the Tamworth Breeders' Club

The Welsh

The Welsh Pig Society was formed in 1922, and the commercial benefits of the breed were promoted, which led to a steep increase in numbers. In 1950 Landrace lines were introduced, and the government identified the Welsh as one of the three significant breeds to be used in the British pig industry. Currently there are 523 Welsh pigs registered; however, sadly the breed is on the RBST 'At Risk' register.

The Welsh pig is a rare breed native to Wales, hardy and an ideal smallholder's pig. Welsh sows are wonderful, milky mothers, producing large litters. The Welsh pig is placid and easy to handle, white in colour, with lop ears, and pear-shaped with good loins and hams. They are happy indoors but do love to root outside in a well fenced paddock or wood. The Welsh has many positive characteristics: my favourite is their personality, though one downside is that they are prone to sunburn, so stock up on suncream to prevent crispy ears in the summer!

A fine Tamworth sow.

A lovely Welsh sow from Hideaway Farm.

Because the Welsh produces excellent bacon, sausages and pork cuts the piglets are easy to sell to grow on. They reach porker weight at roughly twenty-one weeks, so feed conversion is efficient; the pork has just the right amount of fat, and the flavour is sweet and the meat succulent.

Corinna Taylor , Hideaway Farm Pedigree Pigs

ASSURANCE SCHEMES, INCLUDING ORGANIC PIG REARING

In this label-loving world, it can be tempting to do the same for our pigs. Therefore it is really important to understand what you can and can't say about your produce, and the steps you need to take to 'certify' the holding or the meat.

Organic Pig Production

Organic production aims to use natural substances and processes to minimise the environmental impacts of food production, as well as meet high standards of animal welfare.

The term 'organic' is protected under EU law, meaning that to label a holding or a product as organic you must meet very specific regulations as stipulated by the EU's Organic Regulations.[5] This can only be achieved with the certification of an organic assurance scheme, which will check that you have met the requirements, the most popular of which in the UK is the Soil Association. It is really important to note that you cannot call yourself organic without going through the process of conversion (generally taking two to three years) and without meeting every part of the EU's Organic Regulations. Stipulations will be detailed and far reaching, from the feed you provide, the way you rotate land, and the way that you use veterinary medicines. Pigs are required to be kept outdoors (unless in the final finishing stage), and are only weaned after forty days, which is much later than most other commercial holdings. If you are interested, have a look at the scheme standards for an organic certification body such as the Soil Association.[6]

Red Tractor Certified Standards[7]

The Red Tractor label signifies that a holding has reached a base level of standards so the produce is suitable to be sold by supermarkets in the UK. Standards broadly reflect the Code of Practice, with the extra stipulation that pigs must not be castrated under Red Tractor. The scheme allows a mechanism for checking that the food in our supermarkets is produced to an acceptable standard. For pig producers, that includes a three-monthly visit from their vet to check the pigs and review their health and welfare plan, as well as a yearly audit from a Red Tractor inspector. This three-monthly vet visit means that producers and vets develop very close relationships, and vets understand these units thoroughly from seeing them both routinely and when things go wrong.

Most smallholders selling produce locally or to farm shops and butchers will not be required to show Red Tractor certification. But for those wanting greater support and more checks to ensure that they are meeting the standard, Red Tractor offers a sensible set of guidelines to follow. It is important to note that pork imported to the UK (about 60 per cent of the pork that we eat[3]) does not have to meet the same standards, and often falls far short of them.

RSPCA Freedom Foods[8]

This assurance body gives an extra layer of certification and allows produce to be labelled with the 'RSPCA Assured' label. It can be thought of as an enhanced Red Tractor. It is written with the Five Freedoms[2] in mind and has extra stipulations around welfare, such as greater space requirements for all pigs. This is especially pertinent for sows during the suckling period; traditional farrowing crates that limit sow movement to reduce the likelihood of piglets being squashed are not allowed under the scheme. The scheme is a good one for smallholders as it ensures that producers are providing a product at least in line with our higher welfare supermarket producers, who will be required to have both Red Tractor and RSPCA Assured certification.

The Wholesome Food Association[9]

This scheme markets itself as a low-cost alternative to organic certification with a similar intent to raise natural produce using sustainable, non-polluting methods. They don't have scheme standards but instead have a set of principles that producers pledge to adhere to. This scheme is built on trust and therefore producers are not inspected but have an 'open gate' policy allowing consumers to visit the holding, for example through a yearly open day. Producers are then able to label produce with the Wholesome Food Association symbol.

Sue and Stephen's beautiful pork, labelled with care.

Other Labelling

Outdoor pigs can feature extra labelling if they meet certain requirements, as described in the Code of Practice for the Labelling of Pork and Pork Products.[9] It is important that producers only call or label produce with these terms if they are sure that they meet these descriptions. This is so that consumers can be sure of what they are buying, allowing them to make informed choices about their food. More information can be found on the Pork Provenance website,[9] where producers can sign up for this voluntary Code of Practice.

Free range: Pigs are born outside in fields where they remain until they are sent for slaughter. They are free to roam in very generous space allowances within defined boundaries, with good rotational practices. Breeding sows are kept outside in fields with generous minimum space allowances.

Outdoor bred: Pigs are born outside in fields where they are kept until weaning. Breeding sows are kept outside in fields with generous minimum space allowances. Pork and pork products labelled as 'outdoor bred' will also contain a statement about how the weaned pigs are subsequently farmed.

Outdoor reared: Pigs are born outside in fields, where they are reared for approximately half their life (defined as having attained at least 30kg). Breeding sows are kept outside in fields. All pigs are provided with generous minimum space allowances. Pork and pork products labelled as 'outdoor reared' will also contain a statement about the way the grower pigs are then subsequently farmed.

Assurance schemes tend to come with a cost that you may decide is not warranted for you, especially if you would sell produce regardless of this labelling. However, we do still encourage you to look at the assurance scheme that best aligns with your intent, and to refer to and act to meet that scheme's standards. Yes, you won't be able to use their labelling, but you can be more content that you are providing the life for your pigs that you envisaged, which for many is to go above and beyond that of the pig products you can buy in a supermarket.

When consulting scheme standards, be sure that you are consulting the most up-to-date document (which may no longer be the documents on the reference list at the back of this book) as standards change frequently.

SOCIETY MEMBERSHIPS

Networks are hugely important to smallholders; they allow pig keepers to see how it is all put into practice.

Societies for All

Smallholding Vet Schemes

If you aren't already registered with a livestock vet, try to find one with enthusiasm for pigs, and preferably one that runs a smallholder scheme. A smallholder scheme normally means that a practice will have vets dedicated to it with a specific interest in meeting the needs of smallholders. These needs can be quite different from commercial clients, and it generally means that smallholders and vets will be on a similar page. Pete's practice, Synergy Farm Health, has a thriving smallholder scheme that provides regular meetings, a newsletter and health planning as part of its membership.

Pete running a smallholder pig meeting in his younger days.

Social Media Groups

There are hundreds of social media groups that pig keepers can join. Generally, each breed will have one, as will other groups, such as minipig keepers. Some are excellent for posting useful tips and resources, and there are often some really experienced and knowledgeable keepers who might be able to answer questions. Conversely, however, please remember that anyone can provide their opinion on these groups, and therefore it is really important to verify what you read. We frequently see illegal, unethical and inappropriate suggestions and practices on these groups due to their lack of regulation. Some group admins are excellent at quashing these, however others can descend into chaos.

Please also remember that only your vet should be consulted for veterinary advice. It is actually illegal for anyone other than a vet to diagnose a disease in an animal,[10] and advice on the giving of medicines is only allowed to be given by a vet. Responsible keepers may say things such as 'when my pig did a similar thing my vet said "X" and we gave "Y", so call your vet to see if they think the same.' Anything other than this is not appropriate. It is often impossible for a vet to diagnose a condition from a photo or video and suggest the most suitable medication. So please don't trust others with the health of your pigs without that backing.

Breed Societies
All the breeds have enthusiastic societies where keepers can seek mentorship, get to know others keeping the breed, and support its preservation. Sue and Stephen have had a vast amount of support from the Large Black Pig Breeders' Club, and highly recommend them.

Societies for Pig Producers

British Pig Association (BPA)
The BPA runs the herd books for fourteen breeds (all the rarer or less commercial breeds, other than the British Lop and the Kunekune), and they complete conservation work in conjunction with the Rare Breeds Survival Trust (RBST). There are also benefits such as receiving the excellent *Practical Pigs* magazine. Membership is graduated - for example, those just fattening pedigree weaners need only get a 'Pedigree Pork Fattener' membership. Membership gives you access to marketing material and a place on the on-line directory of Pedigree Pork producers. They also automatically register your herd on Defra's 'UK Breeds at Risk' (BAR) register, meaning that in the event of a cull for notifiable disease, your pigs may be spared in order to save the breed.

Societies for Pig Producers in Wales
Porc Blasus: It is free to join this organisation, which promotes pork in Wales. Joining gets you an entry on their on-line map of producers, and regular deliveries of promotional material such as recipe cards.
Menter Moch Cymru: If you intend to make your pigs into a business and are in Wales, then Menter Moch Cymru can offer a great deal of help, such as mentoring your business, free signs, workshops and training.

CHAPTER 2

LEGISLATION AND REGULATION AROUND UK PIG KEEPING

Because all pigs, even pets, are classed in law as food producing animals, the legislation around keeping them is much more detailed than for keeping a dog or a cat. Inspectors can and do check that keepers are adhering to the rules, and ignorance is not an excuse. Legislation can change, so familiarise yourself with the relevant legislation for your area before starting out.

Register the land:[1] Before pigs can be placed on any holding the land must be registered by obtaining a CPH (county, parish, holding) number in England, Wales and Scotland and a holding code in Northern Ireland. These are obtained from the Rural Payments Agency, Rural Payments Wales, Rural Payments and Services, or DAERA, respectively.

Register the herd:[2] Within thirty days of pigs arriving on a holding, the herd must be registered by obtaining a herd number/mark from APHA or DAERA. It is much easier to have this before getting the pigs.

Movements and transport:[2,3] Pigs are only allowed to be moved under licence, and this should be in place before pigs are moved. It is the responsibility of the registered keeper to obtain the licence. Once pigs arrive at their destination, the move and number of pigs received should be confirmed by the new keeper. This is done through electronic systems in England and Wales (eAML2) and Scotland (ScotEID), or paper records from DAERA in Northern Ireland.

Pigs in transit must be accompanied by the movement documents, which will be emailed to you and should be printed. Animal transport certificates, which record the details of the transport, should also be kept. This is all that is required for non-economic movements or economic movements less than 65km. Pigs must not travel if they are not fit for transport (*see* Chapter 7, Transporting Pigs); however, exemptions apply for moving pigs for emergency veterinary treatment.

Pigs on and off the holding:[2,4] A holding register should be documented for all movement of pigs on or off a holding within thirty-six hours of the event. This is automatically completed using the online eAML2 system; alternatively paper or computer records can be used detailing the following:

- The date of the movement
- The herd mark of the pigs being moved
- The number of pigs being moved, and individual identification where necessary for the move
- CPH and details of the holding that pigs are being moved from
- CPH and details of the holding that pigs are being moved to

A yearly inventory of the maximum number of pigs normally present on the holding must be taken and stored on paper or electronically. APHA must be notified if a keeper stops keeping pigs, and records should be kept for three years after someone stops keeping pigs.

Standstill periods:[4] The arrival of pigs on a holding means that no pigs can then leave that holding for twenty days, and no cattle, sheep or goats can leave for six days (thirteen in Scotland). This is intended to prevent the spread of infectious disease. There are some exemptions to this, for example pigs moving directly to slaughter. Showing pigs moved into a Defra-approved isolation facility twenty days before the first show and twenty days after the final show would not trigger standstill for the rest of the holding.

Identification:[2] Pigs aged twelve months or less can move between two farms without a form of permanent identification showing the herd mark, and instead can be identified with temporary identification such as a spray mark. All other pigs being moved for all other instances must carry a form of permanent identification. This can be an ear tag, a tattoo or two slap marks, one on each shoulder, in permanent ink. For showing, exhibiting, moving pigs to a breeding location or for export, identification must also consist of a unique identification number.

Feeding:[5] It is illegal to feed pigs meat, meat products or kitchen scraps. This includes waste from any kitchen producing vegan food. Previously, vegan domestic kitchens were able to feed kitchen scraps to pigs, but recently this has been changed to be illegal as well. We must also ensure that pigs or anything they will be in contact with are not exposed to any meat or meat products, such as dog food (especially raw). Waste milk can be fed if produced from the same holding, but the possibility of transmitting disease and antibiotic residues should be considered. Eggs are not allowed to be fed. A likely entry for the notifiable disease African swine fever (ASF) into the UK is via

It is illegal to feed kitchen scraps to pigs.

contaminated food (for example uncooked pork from an infected animal) being eaten by pigs in the UK.[6]

Medicines:[7,8] Pigs should be registered with a vet. For food-producing animals, records of all veterinary medicinal products purchased, administered and disposed of must be kept. This includes medicines given by the vet. The easiest way of doing this for many smallholders will be to use AHDB's electronic medicines book (eMB).[9] Also adequate are paper records that detail the product and batch number, date of administration, dose, withdrawal and the identification of the pig (temporary identification marks must be visible for the duration of the withdrawal period). These records must be kept for five years. We will explain this in more detail in Chapter 10, 'Medicines'.

Welfare:[7] All keepers of pigs are required by law to have a copy of, and be familiar with the Code of Practice for the welfare of pigs.[10] This has recently been updated for England and Wales and is designed to clarify and expand on legislation around pig keeping. As well as frequently referring to the Code of Practice, we will act to expand on it to enable you to fulfil it. For example, the requirement for keepers to be familiar with the signs associated with notifiable disease and notify the APHA or consult a veterinary surgeon accordingly[11] will be discussed in Chapter 9, 'Diseases of Pigs'.

Walking:[15] If a keeper would like to walk a pig they must apply for an APHA (or AHVLA) licence, which will seek to approve a route. This will involve checking for the presence of commercial pig farms, food-processing premises and livestock markets. Once set, this route cannot be deviated from. The pig will be required to be identified with an ear tag or tattoo, and the licence must be renewed annually.

Sick and deceased pigs:[7, 12] The Pig Veterinary Society Casualty Pig document[13] (which is freely available) should be consulted to aid decision making in cases of casualty pigs and to alleviate suffering. Deceased pigs must be removed as soon as possible, and must not be buried. They can be temporarily stored in a leak-proof bin, but then must be collected by an approved transporter such as a knacker or hunt kennel. They can only be incinerated on farm using a licensed facility.

Land management: To keep pigs on rural land, you may need to apply for an environmental impact assessment (EIA) as covered by EIA regulations.[14] This is outside the scope of this book, but you must make sure that your decision to keep pigs will not adversely affect our rural land. You must also follow regulations on management and disposal of waste, such as slurry, for which the NFU's best practice checklist is helpful.[15]

Be sure to meet the extra legislative requirements for selling produce.

Selling pig meat and food hygiene: There are extra laws around food hygiene for those slaughtering pigs to sell meat, rather than for their own consumption, and keepers will require extra inspections.[16] This is outside the scope of this book (aimed at pig keeping rather than selling meat) and therefore it is up to you to ensure you meet the correct requirements for selling produce.

CHAPTER 3

BIOSECURITY

The Code of Practice describes biosecurity as 'a set of management actions and physical measures designed to reduce the risk of introduction, establishment and spread of disease to, from and within the pig herd'.[1]

Many smallholders aim for a natural level of immunity and don't like to obsess over biosecurity practices. However, it is important to note that the introduction of disease on to a holding, for example by buying a pig carrying an infectious disease, is actually quite unnatural and can have very severe consequences. As an example, many keepers who have inadvertently brought lice on to the holding, thought of as a fairly mild disease in the plethora of available options, will never be able to eradicate it fully, leading to whole herd injections every few months. This is far from natural! Therefore we encourage smallholders to take biosecurity seriously. We agree: your own pigs' dirt is good, but sudden exposure to high levels of disease from other holdings should be avoided.

MOVING PIGS ON TO A HOLDING

To prevent the introduction of disease on to a holding, most commercial pig-breeding farms operate a closed herd. This means that no new pigs come on site, and new bloodlines enter in the form of tubes of boar semen (which is tested for infectious disease). Many will have operated this system for seventy years or more, and an élite few will genuinely be free from our most worrisome pig diseases. The rest will just be stable in their own diseases and no longer suffer badly from them.

Most smallholders, however, will require pigs to enter the holding, and therefore an isolation (quarantine) area for new pigs becomes really important. For those showing pigs, this is crucial. Correct utilisation of an APHA-approved isolation area means that your entire holding does not need to go into standstill every time you return pigs from a show. Remember that without this, when a pig enters a holding, no pigs must move off the holding for the next twenty days.

New pigs entering a holding, or pigs being reintroduced on to a holding, should spend at least two weeks in an isolation area before mixing with the rest of the herd. This is so that the pigs can be closely monitored for signs of disease before being introduced to your pigs.

An isolation area should be on a hard standing, so that it can be thoroughly cleaned and disinfected between uses, and at least three metres away from other pigs. Isolation areas should have dedicated personal protective equipment in the form of a boiler suit, wellies and gloves. At the very least, attend to the area last in the day, and thoroughly clean and disinfect boots and clothing afterwards.

HUMAN CONTACT

Remember to maintain as much human contact with isolated pigs as possible, especially brought in gilts, so that they remain, or become, easy to handle.

Monitor isolated pigs closely for signs of disease such as coughing, diarrhoea, lameness, itching and difficulty breathing. If you see any of these symptoms, consult your veterinary surgeon before you admit these pigs into the main herd. This is one occasion where in-depth diagnostics, or return of the pigs, is warranted to avoid putting your own pigs at risk. If you see signs in one pig in the group, do not introduce any of the pigs until you have got to the bottom of the problem. There are several diseases for which pigs remain a carrier even after they have been successfully treated for the disease, therefore think very carefully before introducing a pig to your herd that has shown signs of disease, even after it has recovered.

The commonest complicating factor to the good practice of isolating pigs is the requirement for pigs to have the company of other pigs. This should only be relevant for breeding boars being loaned or returned. In an ideal world, all loan boars would pair with a castrated friend. If this isn't possible, keep hired boars in the next paddock from some other pigs. Choose a suitable group carefully, and ensure that the boar is secure so he can't pick a fight or breed with any of them. Don't allow nose-to-nose contact, but allow them to have a chat.

Sue and Stephen's isolation area, on a hard standing away from other pigs.

Before buying, verify the vaccination status of any incoming pigs. During the isolation period, make sure that the vaccines that incoming pigs have received are the same as those yours have received, and that incoming pigs will be covered from the diseases that you have on the holding. Vaccination courses often take more than two weeks, so it may be preferable to ask the previous holding to vaccinate the pigs for you. You should worm incoming pigs twice with an ivermectin wormer, two weeks apart. This will kill external and internal parasites, especially lice, which are a real nuisance when brought on to a holding. We discuss this more in Chapter 12, 'Parasite Control'.

SWINE DYSENTERY

The disease swine dysentery is important to mention for smallholdings in this context. This bacterial disease causes diarrhoea (often bloody, but not always) and weight loss (or lack of weight gain). Pigs can die from dysentery, but they can also carry the disease with few clinical signs, infecting other pigs that they come across. The disease is devastating for holdings because it is so difficult to eradicate once it has taken hold. Therefore, preventing swine dysentery warrants particular attention when moving pigs on and off holdings. Show pigs have been highlighted as a specific risk for the spread of swine dysentery. The disease is so serious that holdings with it are encouraged to log this on AHDB's 'significant diseases charter'[2] so that we can understand where it is in the country.

Where isolation is not possible, you must be able to trust the source of incoming stock. This includes the holdings to which you loan your boar. Gain a thorough disease history from the holding, especially for parasites (internal and external), diarrhoea and reproductive problems. Be sure to ask for courses of vaccines and wormers to be completed before pigs arrive.

For those simply rearing weaners to go to slaughter, best practice may be more easily replicated. Commercial pig farms rearing pigs (often taking weaners through to finishing) operate what we call an all-in-all-out (AIAO) system in which the unit is emptied of pigs, fully cleaned, disinfected and dried before it is filled with pigs again. In this way, those buying yearly weaners can clean and disinfect housing and leave a paddock free from pigs for the next few months. Most pig pathogens will die in the soil after a few months so this should do a similar job to an AIAO system, especially over a cold winter. If the last group of piglets experienced diarrhoea, be sure to use a disinfectant that will kill coccidia.

MOVEMENT OF PEOPLE AND OBJECTS ON AND OFF THE HOLDING

Whilst most diseases enter through incoming pigs, it is also important to consider incoming people and objects. Pathogens can sit in the tread of a wellington boot or in the tyre of a vehicle, and it may mean that you bring home more than you think. A line of separation between off farm and on farm allows you only to allow visitors and vehicles that you trust past this point. AHDB provides signage to make this clear to visitors.[3] Collection of carcasses should take place, where possible, at the perimeter so that external vehicles do not enter the holding. A boot dip filled with clean disinfectant costs very little, gives an extra line of defence, and also lets incoming visitors know that you are serious about preventing disease on your holding.

Some of Pete's smallholders proudly displaying their AHDB biosecurity signs.

Sue and Stephen's diligently completed visitors' book.

Ask that visitors wear clean clothes and wellies after being with other pigs, and where possible, ask that visitors haven't been near their pigs that day. Any visitors to the holding should sign a visitors' book. This is so that notifiable disease can be traced if it occurred. Red Tractor states that this should display:[4]

- the date of the visit
- the name of the visitor and the organisation if relevant
- the purpose of the visit
- the date of the last contact with pigs
- confirmation that the person hasn't experienced vomiting, diarrhoea or flu-like symptoms in the last twenty-four hours

OTHER PRECAUTIONS

Movement Within the Holding

On the holding, different aged pigs should be kept separate where possible. It is best practice to move from the youngest piglets to the oldest pigs through the day, as piglets are less resilient to disease. Pay particular attention when moving from a sick pig that might have an infectious disease back to the main holding, and clean and disinfect where appropriate.

Human Safety

As covered in Chapter 9, 'Diseases of Pigs', pigs can transmit diseases to humans, and vice versa. Wash your hands thoroughly after tending to pigs, especially before eating, and wear gloves if they are showing any signs of disease. Pigs and humans can catch the same type of flu virus, so if you have flu, don't go near pigs as you may infect them.

Notifiable Disease

Be sure never to bring pork products on to the holding to ensure that notifiable disease, such as African Swine Fever, is not passed to your pigs. This would lead to devastating effects, including culls. Members of the public may be less aware of suitable behaviour around pigs. Signage, which can be provided by AHDB,[3] should display the risk of feeding pigs.

Pests and Wildlife

Rodents can be carriers of diseases to pigs, such as salmonella, swine dysentery and leptospirosis. Therefore a rodent control plan should be enacted. Wild boar are often less affected by disease than our domestic pigs, and therefore do not have to appear sick to cause a problem. Among other diseases, wild boar have been implicated in the spread of African Swine Fever in Europe. They will visit for food and a date with your sows, so must be allowed no access to either of these.

USE OF DISINFECTANTS

Suitable disinfectant will depend on the pathogens you want to kill. The table includes some commonly used disinfectants on farms, and the suitable concentration for general use of these disinfectants. Please note that other disinfectants are available, but you must use a Defra-approved disinfectant at the correct concentration (dilution rate). Also note that this dilution rate will be different in the case of notifiable disease, such as foot-and-mouth disease

Some commonly used disinfectants with Defra's recommendations of suitable concentrations for general use[5]

Product	Company Name	Concentration
FAM 30 Evans	Vanodine International Plc	1 part disinfectant to 49 parts water
Bi-OO-cyst	Biolink Ltd	1 part disinfectant to 50 parts water
Iodo Pharm	East Riding Farm Services Ltd	1 part disinfectant to 50 parts water
Total Farm Disinfectant	Downland Marketing Ltd	1 part disinfectant to 49 parts water
Virkon LSP	Antec International Limited	1 part disinfectant to 40 parts water

Pete using an appropriate foot dip.

or TB, and the Defra website should be consulted for these revised figures. If you have had an outbreak of coccidiosis or Cryptosporidium (both causes of diarrhoea in piglets), Bi-OO-cyst is the recommended product to use.

When constructing a boot dip, note that many disinfectants are inactivated by organic matter (poo), so a dirty boot dip is likely to be ineffective. Some diseases, such as swine dysentery, can also cling on to a tiny amount of organic matter and remain infectious. Therefore, have one bucket of water with a hoof pick or brush to remove faecal matter, or a powerful hose, followed by a boot dip of disinfectant. To prevent over-dilution of disinfectant (which would reduce its effectiveness) boot dips should have a lid.

BIOSECURITY AT SHOWS

For many, the show season is the highlight of their pig-keeping calendar. It is important for maintaining networks of enthusiasts, and it is likely that precious lines would have died out without the British pig-showing community.

However, moving pigs around the country and mixing them with other herds several times a season is likely to expose them, and consequently your other pigs, to disease. We know from the Covid-19 outbreak that individuals can be infectious to others whilst not showing symptoms of disease themselves. Therefore the only way to ensure that your pigs do not contract infectious disease is to not take them showing and to operate a closed herd.

If you are still passionate that you would like to show pigs, here are some tips to minimise the risk:

- Use excellent biosecurity throughout the season.
- Disinfect, disinfect and disinfect again. Your trailer, your pen that you move into, your equipment, even yourself. Pathogens can last on surfaces for months. Remember that organic matter deactivates many disinfectants, so remove dirt from surfaces and then disinfect.
- Ensure that you have an APHA-approved isolation facility, and use it properly.

The show ring at the Dorset County show.

- If you hear of trouble, don't go.
- If there are rumours and underground chatter, don't go.
- If you see any signs of disease in your pigs, stop the season. We understand that this is heart-breaking and that you may have worked your entire year for this. However, we are now used to the sad fact of having to miss out if we get sick. Let's give our pigs that common courtesy, too.
- If you know that you have an infectious disease on the holding that could pass to other holdings at a show, you should not be showing pigs.
- Know the diseases to watch out for (*see* box).
- Educate those around you.
- Discuss your worries with others, and the steps that you take to minimise the risks. If you see signs of disease in pigs at a show, speak to someone. If you do not feel comfortable speaking to the owner of the pig, speak to the show committee.

AN APHA-APPROVED ISOLATION FACILITY

An APHA-approved isolation facility allows you to return your show team into an isolation area of the holding without triggering standstill on your other pigs. Pigs should move in twenty days before the first show and remain for twenty days after the final show. During the twenty-day standstill period after a show, monitor pigs closely for signs of disease. Consult your vet if you see any clinical signs. Do not give medicine without finding out what the problem is, as medication may well mask the disease without curing it. Whilst in isolation, worm your pigs with an ivermectin twice, two weeks apart, to kill any lice or mange (including egg stages).

RECOGNISE THE SIGNS OF INFECTIOUS DISEASES

Know the signs of infectious diseases such as swine dysentery, atrophic rhinitis, and notifiable diseases. Wasting and diarrhoea in show pigs should be treated as dysentery until proven otherwise.
If classical swine fever (a more chronic swine fever and a notifiable disease) were to creep into the showing community, insidiously throughout the show season, the whole country would very quickly be affected and hundreds of bloodlines lost.

CHAPTER 4

PIG HOUSING AND ENVIRONMENT

Pig housing and environment must satisfy the Five Freedoms,[1] such as the need to provide pigs with freedom from discomfort and disease, as well as freedom to express normal behaviour.

THE BIG QUESTION: INSIDE OR OUT

Most smallholders will have an image of outdoor pigs, however there are pros and cons to both indoor and outdoor systems.

A well organised outdoor smallholding.

Inside, pigs take very much less room, avoid mud and the need for pig arcs. Inside systems allow for closer monitoring, which is especially pertinent for farrowing. On the other hand, more bedding will be required, with regular cleaning out to avoid pigs getting sick. Housing must be well designed to allow for good ventilation but also prevent chilling in winter. Mixed-age housing will also be more prone to disease spreading from older to younger pigs. When housing pigs inside, lighting must be available to inspect the pigs at any time.[2] Inside, expression of natural behaviours will be reduced, but can be assisted in the provision of 'enrichment' such as straw to root around in (on top of that provided for bedding) or horse-feed balls to minimise boredom.

Outside, pigs require more space, and mud can be a problem. There are sites with good drainage where mud can be strategically overcome year round with rotation, paddock rest and use of hard standings to avoid poaching, but this is not the norm in the UK. Please do consider that winter can be a miserable time for pigs. Despite their reputation, pigs are clean animals with their own designated toilet area, which is typically in the coolest area of their habitat.

Pigs can quickly turn grass cover into mud.

Infrastructurally, strong fencing and arcs will be required. Pig arcs are metal huts that can house pigs. They are sometimes also called arks.

Growth may be less efficient where pigs are using more energy to keep warm. However, pigs can express more natural behaviours: rooting, wallowing, running and playing.

ENVIRONMENTAL FUNDAMENTALS

Water

By law, water must be provided ad lib (as much as is wanted) to all pigs from two weeks of age. Pigs are very sensitive to water deprivation and will very quickly start to show neurological signs where they do not have access to water. This can lead to death in severe cases.

Water should be as clean as possible. Drinking water should be separate from a wallow, and the supply designed in a way that minimises the chances of a pig defecating in it. Water should be situated away from bedding areas to keep bedding clean and dry, but no more than 10m away.[3] Water troughs must avoid electric fence lines, and supply must be checked at least daily. Groups of more than ten pigs should have access to two water points, in case there are disagreements in the group.

WATER CONSUMPTION

A pig will drink two to three times the weight of food that it eats.[4] Particular attention needs to be paid to farrowing sows to ensure that water provision is adequate to maintain milk production. On a warm day, lactating sows drink up to 50ltr of water.[5]

Water can be provided by the following means:

- Nipple drinkers are commonly found on commercial operations for cleanliness reasons. However, simpler designs can lead to interruptions in supply, and pigs are unable to sip water from most designs (which they enjoy). If using nipple drinkers, ensure that flow rate is adequate, even when all pigs drink at once, and that the drinker height is appropriate for the age of the pig.
- Tyre drinkers and other containers are easily available and work well. Just ask at your local garage for an old tyre, and the inserts are readily available from agricultural suppliers. However, pigs from finisher weight or above can overturn them, so they need filling up and checking regularly. They also need cleaning out regularly as they can get very muddy. Other containers can be used similarly, but they must be heavy and stable.
- Plumbed-in troughs are by far the easiest, but still need regular checking and cleaning. These can either be plumbed in to the supply directly, or into an IBC (intermediate bulk container) via a hosepipe, so they needn't be difficult or expensive to fit - just make sure the hose is not within reach of inquisitive noses.

A pig tipping over their container drinker may be a hint to top up the wallow!

A tyre drinker provides a practical and affordable solution for these weaners.

Food

Food must be provided in a way that is accessible to the pigs, rather than trodden into mud. If feeding meals (rather than ad lib) it is really important to ensure that all pigs can access the food at once, so that some pigs don't eat all of it and some have none. Pigs are grazers so feed should be split into multiple feeds per day and wherever possible, there should be vegetational grazing available. Diet constitution is discussed in Chapter 5, 'Feeding' but in terms of feed bowls, there are basically three options: troughs, rubber trugs, or feeding on the ground.

- Troughs are great if you have a number of pigs - four plus, say - but they hold rainwater and can get filled with mud. You will also need to go into the paddock to fill a trough, which could mean trying to walk over deep or very slippery mud whilst being tripped up and pushed by excited young pigs. You will then need to empty the rain from the trough, whilst not spilling the feed. 'Mexican hat' feeders are often recommended as they are hard-wearing and can't be tipped up by the pigs, but they are expensive.

Sue and Stephen feed from rubber trugs.

- Rubber trugs are Sue and Stephen's standard receptacle for feeding. Two or three weaners can share one to start with, then should move on to one each as they grow. These trugs are indestructible and can be dropped into the paddock and then collected from over the fence, avoiding having to go into the enclosure.
- Feeding on the ground is a good way to entertain the pigs and keep them occupied, but is only possible on dry ground, otherwise food is wasted.

Food storage must be rat proof and must keep feed dry. Pigs are very sensitive to mould or 'mycotoxins' developing in feed and bedding. Sue and Stephen find dustbins work well, as do trailers, and for larger amounts plastic pallet boxes are a cheap and effective solution.

Plastic pallet boxes are a cost-effective solution to keeping food dry and inaccessible to rats.

Clean, dry straw is crucial for many aspects of pig keeping.

Bedding

Pigs are clean creatures, and don't foul a bedding area unless the area is too large or there is a draught – but they do love to snuggle into clean, dry straw. Bedding can also be a really important insulator, especially for piglets. Straw should be of good quality, kept dry, and cleaned out regularly. The smallest (generally invisible) bit of mould on bedding can lead to issues with mycotoxicity and a plethora of possible clinical signs that will be covered in Chapter 9, 'Diseases of Pigs'. Wheat straw is a safe option as pigs don't eat lots of it, it doesn't absorb much water, and is suitable for all ages as it doesn't 'wrap' piglets.[3]

Pig Shelters

Pig shelter can be provided by a traditional pig arc, or some kind of stable or sty. Arcs can be commercially bought or home-made, but bear in mind that pigs are very strong and love scratching. Home-made shelters must be solid as, under the Code of Practice,[6] pigs must have a warm, draught-free lying area. The design must allow for good ventilation without draughts, and be easy to clean. Some have floors and some don't - a floor makes the arc much heavier, but stops water running underneath, especially on a slope.

Traditional corrugated-iron arcs are popular and cheap, but can be hot in summer and cold in winter. Recycled plastic arcs are more expensive to buy but are virtually indestructible and last longer than both wood and tin. Modern designs will be insulated to keep the occupants warm in the winter

This custom-built 'pig palace' is where smallholders Luke and Laura Armitage keep their sow and litter in Cumbria, where extreme winters are a challenge. The structure is waterproof and well ventilated, and the pigs have continuous access to the outdoors.

A traditional tin pig arc.

and cool in the summer, and have manually controlled flaps to allow ventilation in the summer. Wooden designs can look attractive, but are very heavy to move.

Pigs love to snuggle up together to sleep, so a 5ft × 4ft arc is fine for two weaners. Arcs should be sited on a flat piece of ground so that no one rolls down a hill. This is especially important for lactating sows to reduce the risk of the sow accidentally squashing piglets. The rest of the paddock can be on a hill (this will actually help drainage), but site arcs on flat land on the high ground (so they don't get waterlogged).

Protection from the Weather

Pigs require protection from the elements and thermal comfort. Piglets are especially susceptible to cold, and all pigs will utilise more energy during the winter months to keep warm. Pigs also struggle hugely in the heat, and in the worst cases, heat stress can lead to death.

Cold Weather

Face arcs away from prevailing weather in the winter (so that the back of the arc faces the worst of the weather, generally north). Dig in the base of arcs to reduce draughts, and cover any pop holes.[3] If using stables, they should have four sides in the winter, and ceilings shouldn't be so high that the heat sits above the pigs. Always make sure that pigs have enough clean, dry bedding, and provide extra bedding in more extreme conditions. In cold weather pigs require energy to keep warm, so will need to eat more (sometimes 10–15 per cent more[4]). Gut fermentation

As long as the basics are provided, piglets are surprisingly hardy when only a few weeks old.

generates heat,[4] so you may find that your pigs eat more bedding in the winter, which will need replacing. As long as these basics are done, cold weather is generally only a problem for very small piglets.

Hot Weather

Pigs particularly suffer from sunburn and heat stroke, partly because they are unable to sweat. Hot temperatures also reduce food intake, which is especially dangerous for lactating sows, who need to be eating as much as possible to produce milk for their piglets.

Move arcs in the summer for maximal airflow, turned 180 degrees from their winter position, with shade at the hottest part of the day. Site paddocks so that there is shade at all times of the day, or hang an awning so that pigs can escape the heat. Paint tin arcs white yearly to reduce their internal temperature, which can decrease it by up to 7°C.[7] Design stables to be adaptable in terms of ventilation, allowing airflow in the summer. In very hot weather, ensure that sows have a cool concrete area to lie on that is not bedded with straw (away from their bedding area). Provide pigs with wallows as well as plenty of fresh, clean water. In some cases you may need to apply sunscreen, but try natural approaches such as wallows and shade first.

A shaded arc where they can shelter throughout the hottest period of the day is keeping these pigs more comfortable.

These piglets are keeping cool under the shade of an awning.

MAKING A WALLOW

The mud in a wallow allows water to evaporate from the skin more slowly, allowing the cooling benefits to persist for longer. Mud is also a natural sunscreen. Wallowing is also enjoyable for pigs, and therefore wallows may be utilised in cooler weather too. Some wild pigs seem to use mud baths to scrape off any parasites such as ticks: the drying mud traps the parasites, making them easier to remove.

To make a wallow, just dig out a shallow depression and pour in water – the pigs will do the rest. Wallows should be wet and muddy, so a natural wallow is much better than a paddling pool. Make sure the wallow is easily accessible for topping up, but not near a gate or arc, to avoid those areas also becoming muddy. Try to make wallows slightly away from a pig's water, or at least not surrounding it. Drinking from the wallow will make pigs more likely to pick up pathogens.

Wallows aren't just for pig producers! This pet pig is enjoying his under an awning.

A shaded wallow for a very warm day.

Freedom from Injury and Disease

Pigs should be situated in an area that allows for monitoring at least once a day. When designing paddocks, consider how pigs would be restrained, loaded on to a trailer, or moved so as to reduce the mud burden but maintain practicality.

Resources should be provided so that all pigs can access them at the same time and are not subject to competition. Splitting pigs by age and size will assist in this, and will also mean that pigs are less likely to pick up disease from older, more resistant animals. Pigs should be separated in terms of sex from weaning, where breeding is not intended.

Good biosecurity should be practised, and appropriate signs should prevent pigs being fed by members of the public. Piglets should be protected from predators such as foxes or buzzards.

Paddocks should be walked frequently to check for the presence of debris, breakages and toxic substances. Common toxicities are coal-tar poisoning, rodenticide toxicity, lead toxicity and plant poisonings.[8] Common plant poisonings include bracken, hemlock and black nightshade. Less common examples are rhododendron plants, foxglove, yew and oleander. Parsnips and parsley can cause skin issues, such as an increased susceptibility to sunburn. Pigs are more resistant to acorn and ragwort toxicity than other farm animals, but these can cause issues if ingested in large volumes.

An old ploughshare that Sue and Stephen's pigs dug up.

Bracken is toxic to pigs.

Hemlock water dropwort is toxic to pigs.

Expression of Natural Behaviour

Pigs are highly sociable animals and should not be housed alone. They must be housed so that natural behaviours can be expressed, such as rooting and scratching. A scratching post can also be a useful barometer of health, as excessive scratching can signify a problem with external parasites.

A lovely scratching post constructed for two pet pigs.

Other Practicalities

If you have a group of adolescent boars known for being escape artists, don't house them very close to females. Likewise, if you are keeping a boar on site, consider how close you site females. Try to ensure that at least one party is firmly secured with stock fencing to prevent accidents.

FARROWING ACCOMMODATION

Farrowing accommodation aims to suit two ages of pig with very different requirements - piglets and sows. A piglet's ideal temperature at birth is 28-32°C,[9] whereas a sow prefers about 16°C.[10] If sows overheat they are more likely to suffer from mastitis (inflammation of the udder) and not produce milk. Conversely, on top of the direct effects of hypothermia, a cold piglet does not seek out food and may die from dehydration and starvation. In addition to achieving the right temperature for both sows and piglets, accommodation design can really impact the likelihood of a piglet being crushed by the sow.

Indoor farrowing allows for vastly better monitoring and reaction to problems throughout the farrowing and lactation process. Inside, the area should be well lit, and a creep area should be provided, which is an area where piglets can keep warm and avoid being crushed.[6] It should be bedded with at least 5cm of shavings or straw, with a non-slip surface. It can be constructed using a converted hurdle in the corner of a pen. A heat lamp should be placed above the creep area so that a temperature of around 30°C can be achieved. This will be necessary on all but the hottest of days. Electricity will be required for the provision of heat lamps, and these should be hung in a way that does not present a fire risk.

This old hurdle had its loops ground off, and is simply dropped behind four ring bolts, then tied down to prevent it being lifted.

This farrowing space looks good, but the heat lamp could do with lowering when piglets arrive.

Indoor farrowing spaces should be at least $5m^2$, with a bedding area of at least $2.8m^2$.[11] This $5m^2$ does not include the creep area, which should be in addition. The creep area should be big enough so that all piglets can lie down (about $1m^2$ up to four weeks of age is sufficient[11]) with extra space for feeding.

Farrowing in arcs relies on the mothering ability of the sow, as this system allows for less monitoring, less chance of safe intervention, and a greater chance of sows lying on and suffocating young piglets. Despite this, across the country thousands of pigs farrow in arcs every year with very few problems. Individual outdoor farrowing accommodation should be $20m^2$ [3] and paddocks should be rotated between litters.

The risk of piglets being crushed can be reduced by siting arcs on flat ground and flattening the bed daily to reduce the chance of a piglet falling into the middle and being squashed. Some farrowing arcs will have rails to try to reduce this risk. Fenders (a metal box that fits on to the front of the arc) should be used during the post-farrowing period to stop piglets straying.[6] Sows can become very aggressive during farrowing or nursing, so don't get into an arc with a farrowing sow.

Check out AHDB Pork's page 'How to manage the farrowing arc or hut'[3] for further information.

Arcs should be specifically designed for farrowing, meaning that they are big enough to allow the sow to fully turn around, nest and nurse comfortably,[3] which is at least $2.5m^2$.[12] A heat lamp is not necessary or practicable. Monitor your set-up to ensure that piglets can spend plenty of time nursing, as the more time spent nursing, the healthier and warmer the piglets will be. Hot weather may cause them to spend more time out of the arc, which is detrimental to milk intake and may lead to weak piglets. In warm weather sows will also eat less, which will be detrimental to milk production. Therefore, keeping arcs cool in the summer is just as important as keeping them warm in the winter. Arcs should be moved on to fresh ground and cleaned and disinfected between litters.[3]

Farrowing accommodation is generally best with sows housed individually. Some sows will be amenable to running with another sow and litter, but each should have its own area (for example, its own arc or stable section) that can be closed off from the other sow and litter when required. The sow should move into her accommodation one week before farrowing, and legally piglets should not be weaned for at least four weeks after birth (unless weaning into specialist accommodation).[6] Most smallholdings will wean between six and eight weeks. Piglets must have ad lib access to water from two weeks of age.[6]

SPACE REQUIREMENTS FOR OUTDOOR PIGS

So how many pigs can I put on my land? Space requirement will vary considerably, and will depend on the following factors:

- the type of grazing
- soil type
- the weather
- water drainage
- vehicle use
- paddock organisation, which will affect how easy it is to graze rotationally

It would therefore be unwise to stipulate exact requirements. That being said, it is useful to have some figures to work from. The Soil Association (an organic farming assurance body) provides maximum stocking rates to comply with maximum nitrate application.[12] Given that the Soil Association demands that pig keepers manage stocking density so that vegetation is not over-grazed and soil is not poached, this is a sensible minimum to work from. The chart does not mean that pigs should be given this total amount of space all at once, but that a rotational system should be used to maintain vegetation cover at all times and reduce mud.

Using the Soil Association's data in the chart in the next column, work out the space required for the maximum number of pigs that you will have on the ground at any one time. Then increase stocking densities very slowly over time to establish what works for you and your holding. Following this, think about individual paddock sizes. In a fresh paddock, free-range growing pigs should have $12m^2$ per pig.[11]

The Soil Association's maximum stocking rates to comply with maximum nitrate applications[12]

Weight	*Maximum stocking rate per hectare*
7-13kg	170
13kg-31kg	40
31kg-66kg	22
66kg, intended for slaughter	16
Breeding sow before her first litter	15
Sow with litter up to 7kg	9
Breeding boar 66kg-150kg	14
Breeding boar over 150kg	10

FENCING

Pigs can be tricky to keep enclosed, meaning that fencing must be carefully designed. Smallholders can choose to use either stock fencing, electric fencing, or a mixture of the two. Stock fencing is strongly advisable, at least around the perimeter of a holding so that escapees cannot go far. The traditional form of stock fencing for pigs has a line of barb at the bottom; however, barb can injure piglets so we recommend just stock fencing.

We also recommend tensioning the fence as much as possible, and putting the posts closer together than normal. Stock fencing can also be used around paddocks; however, using stock fencing means that you can't adjust paddocks as easily, and you should be aware that piglets may still be able to escape under gates. Gates should be metal and mounted so that they can't be lifted from their hinges.

Electric fencing allows greater flexibility in the use of your paddocks and land organisation. The Code of Practice[6] states that electric fences should be set up so that 'when the animals come into contact with them they do not feel more than momentary discomfort'. A pig's natural instinct to pain is to run forwards and therefore through the

Well constructed stock fencing is a safe option for pigs.

fence; to avoid this some training is needed. The recommended method is to set up an electric fence with a standard and obvious fence just behind, so the pigs learn to jump backwards, not forwards.

Alternatively, Sue and Stephen train piglets by spending a couple of hours with them once they go into the electric fenced paddock. Whenever piglets approach the fence Sue and Stephen stand the other side with a pig board blocking them, and the piglets learn very quickly. Even if you don't use electric fencing yourself, if you are raising piglets to sell it is advisable to train them to ensure that they will be well adapted for their next home. Sue and Stephen find that three strands of tape, the middle one roughly at nose height, works

The bar added at the bottom of stock fencing prevents piglets from escaping.

Whatever the design, stock fencing must be very strong.

Piglets are escape artists!

well. Tape is preferable to wire as the pigs can see it more easily.

Site arcs at least 1.5m away from electric fencing,[3] and be absolutely sure that electric fencing is not touching troughs or gates to avoid the risk of pigs and people being electrocuted. Grass will short the circuit, effectively earthing an electric fence, so mow round where you are about to lay it. Some keepers will run a single line of electric tape a short distance inside a stock fence.

The grass is quite long here so is likely to have shorted out the circuit, allowing this little piggy to run free.

Pigs and dogs should never be left unattended.

Beyond this, having good-sized paddocks with plenty of grazing will make pigs unlikely to want to leave their paddocks, and training pigs to come to call means that it is less of an issue if piglets escape but stay within the holding.

HOUSING PET PIGS

There is an increasing trend to keep a 'house' pig – that is, a pig that lives indoors with your family and other pets; however, this is not something we recommend. House pigs can become problematic where the expression of natural behaviours is seen as destructive or aggressive. This must not be discouraged, and instead suitable outlets must be sought. Ensuring legislative compliance, such as not feeding anything from a human kitchen, can also be difficult to achieve if the pig is kept indoors.

Be very careful if pigs and dogs are allowed to share space, and never leave them unattended together. Pigs and dogs often struggle to read each other's body language, and can fight unpredictably and dangerously.

CHAPTER 5

FEEDING

THE CONSTITUENTS OF A PIG DIET

There are several 'key ingredients' necessary to pigs, as itemised in the following section.

Water: Water is crucial for life, but luckily is simple to provide.

Fats, oils and carbohydrates: Energy is required for all bodily processes, such as keeping warm, moving around, or for cellular reactions. Energy is delivered in the form of fats, oils and carbohydrates.

Fatty acids: Fatty acids are the building blocks of fat and are made up of essential fatty acids and non-essential fatty acids. Essential fatty acids (polyunsaturated fats such as omega-3 or omega-6) are those that cannot be synthesised by the pig and therefore must be delivered in their feed.[1] A pig's 'fullness', or its decision to stop eating, is mostly determined by its energy needs being met[2] through the hormone leptin.[3]

Protein: Protein is made up of chains of amino acids, of which around twenty are naturally occurring. Different orders of these amino acids are needed to form different tissues, hormones, antibodies and more. Ten of these amino acids cannot be synthesised by the pig and are therefore termed 'essential amino acids'. By this, we mean that it is essential that these are contained in the diet. The need for essential amino acids is highest (as a proportion of total diet) in young piglets, which require a high level of protein to grow. This decreases as the pig ages. Protein-rich foods include peas, beans and other pulses. When thinking about providing protein, remember that it is illegal to feed protein that is meat derived.[4] This also means that pigs cannot be given eggs from the holding.

The 'essential amino acids' that pigs require the most of are lysine and, to a lesser extent, methionine, and others. Therefore keepers should look out for information on the inclusion of these essential amino acids, which is printed on feed labels. It is important to note that the quantity of 'crude protein' in a food means very little without scrutiny of the inclusion of these essential amino acids. A protein-rich ingredient such as soya is actually low in lysine as a proportion of its total protein, meaning the pig would need to eat large quantities of soya to meet their lysine needs.

Excess amino acids cannot be stored in the body and are therefore excreted in the urine, so feeding too much of other amino acids is harmful to the environment (due to their nitrogen composition) and represents a waste of feed and money. To avoid this imbalance, some food companies add neat forms of these essential amino acids.[5]

Vitamins and minerals: These are crucial to avoid deficiencies, especially in young piglets, but are only needed in small quantities. Feed companies will add these to commercial feed, and advice from a professional nutritionist

A typical feed label showing constituents.

must be sought if keepers would like to investigate adding these themselves. The balance of vitamins and minerals is critical to avoid disease, and requirements will be different for different stages of physical development and pregnancy.

Many vitamins and minerals work together in the diet. Pigs at pasture are able to ingest some vitamins and minerals from the soil, but this is hugely variable from one holding to the next so can't be relied upon.[6]

THE BALANCE OF VITAMINS AND MINERALS IS CRITICAL

Calcium, phosphorus and vitamin D will lead to bone disorders if not given in the correct ratios,[7] with consequences resulting from both over- and underdosing.

Vitamin E (selenium) deficiency leads to sudden death, most commonly in piglets soon after weaning.[7] Commercially reared piglets are fed specific diets that are precisely supplemented for each life stage, meaning that that condition is now only rarely seen. However, mulberry heart disease should be considered in cases of sudden death of piglets on smallholdings.

Zinc has long been added at high levels to commercial growing piglet diets to reduce the occurrence of post-weaning diarrhoea.[8] This is in the process of being made illegal due to concerns of the effect of heavy metals on our environment, and therefore alternative strategies must be sought. It requires a veterinary prescription, so you will know if you will be affected by this ban.

Fibre: Fibre or roughage is beneficial for immunological gut function and will fill the stomach, making an animal feel fuller.[9] Therefore, fibre can be used where an animal may be at risk of becoming fat, such as pet pigs or dry sows. Fibre does provide a small amount of energy as it is digested by gut micro-organisms and then absorbed by the pig.

Unwanted substances: Pigs are especially sensitive to toxins that can be found in bedding and feed, which can cause a huge range of clinical signs from vomiting to infertility.[10] 'Mycotoxins' are produced by mould and fungi. They can be visible on feed or bedding but can also be invisible. They like moist, warm environments[10] so it is imperative that feed is dry and kept hygienically. The pig diet must also be free of toxic plants.

Mouldy feed that should not be fed to pigs!

ACHIEVING A BALANCE

The constituents of a pig's diet need to be finely balanced to ensure that the pig receives everything it needs. Going outside this balance will lead to food waste and inefficient feeding at best, and deficiencies or stunting at worst. We can monitor the suitability of diets by looking at growth rates and fat levels, faecal consistency and overall health.

The balance of the diet constituents required will be different in rare breeds to commercial pigs, whose genotype has been 'improved' for maximal lean growth. In general, rare-breed pigs require lower protein and higher energy feed.[6] If a pig receives too much energy compared to the level of protein, it will become fat.

Balancing the amount of food is also crucial for pigs. Emaciated or over-fat pigs can experience disease problems, for example joint disease, they will farrow less successfully, and will not yield good produce. Body-condition scoring allows us to assess objectively whether our pigs are too fat or too thin. It is important to body-condition score your pigs frequently and act on this information. The body-condition score is graded from 1 to 5 in pigs, with 3 being ideal in the vast majority of cases. Body-condition scoring is done by feeling bony prominences, and judging how easily bones can be felt or seen. Scoring from behind a pig is useful as it isn't affected by pregnancy and lactation, but assessing the shoulders and ribs is also important.

Nutrition should be increased or decreased gradually to achieve the correct score for that pig. Don't guess the weight of food: invest in a good pair of scales and use them every

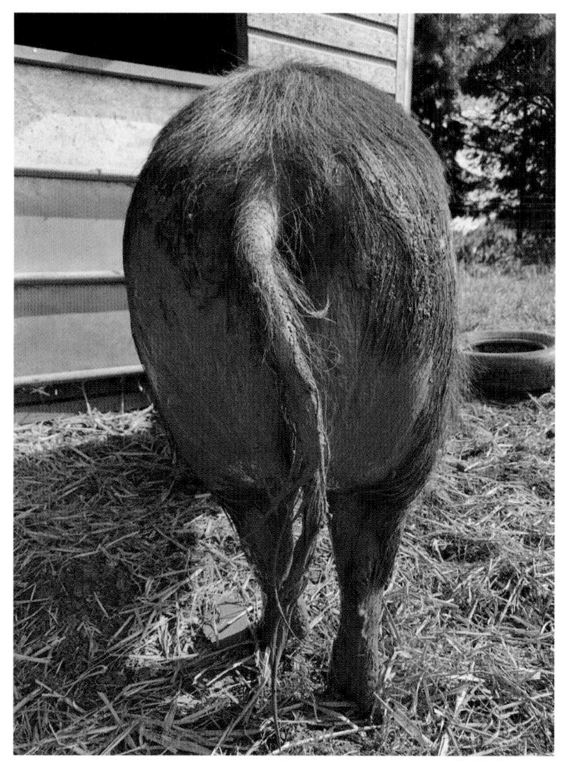

This is the perfect angle to body-condition score a pig. We would call this pig a 3.5, and she would be in a good state to start lactation. Note the slight fatty swelling around her tail base.

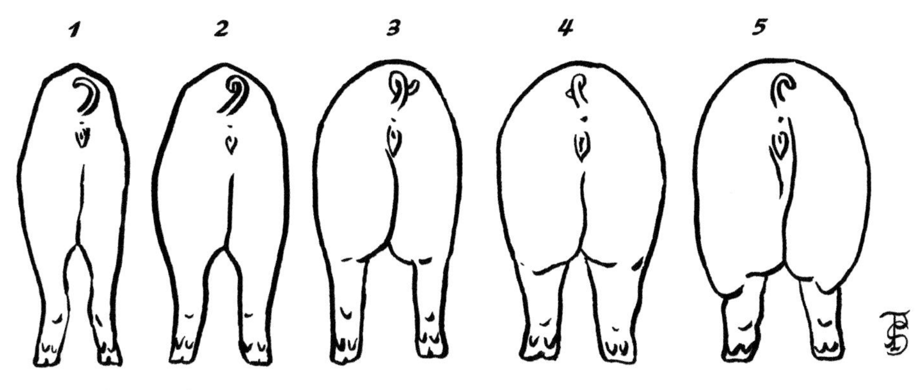

BCS scoring values for pigs:[1]

1 is an emaciated pig, where 'shoulders, individual ribs, hips and backbone are visually apparent'.
2 is a thin pig, where 'shoulders, ribs, hips and backbone are quite easily felt when pressure is applied with the palm of the hand'.
3 is optimal, where 'shoulders, ribs, hips and backbone can only be felt when pressure is applied'.
4 is fat, where 'shoulders, ribs, hips and backbone cannot be felt even when pressure is applied'.
5 is grossly fat, where 'fat deposits are clearly visible'.

We would call this a body-condition score 3, which would be perfect to finish her lactation.

A well organised feeding station allows Sue and Stephen to manage pig feeding successfully.

Weighing out food every time prevents pigs being fed too much.

time you feed. Consider competition between pigs to ensure that the intended diet is getting into the intended pig. Bullying can be especially apparent where concentrate feed provisions are minimal, for example with pet pigs or dry sows. If you have pigs that are not gaining weight as you would expect, please contact your vet.

CHOICE OF DIET

Commercially Produced Diets

The complexities discussed mean that it is generally most appropriate for smallholders to feed a commercially produced compound feed. Commercial diets come in different particle sizes:

- 'Meal' diets are the diet in their natural, powdery form. Meal increases the possibility of food waste and vermin, as pigs tend to throw it all over the floor. Meal diets do, however, slow down gut transit time and can therefore be helpful in pigs experiencing scours.[11]

An example of pig pellets: these are sow rolls.

An example of pig pellets: these are sow pencils.

- Pelleted diets are more frequently fed, where the meal is formulated into a pellet and coated.[11] The pellet size can vary from small weaner pellets right up to large sow rolls. Sow rolls are useful if feeding on the floor because they decrease waste, as pigs can find larger pellets more easily. To achieve this particle size, the feed has to be bashed to be very small. This can cause stomach ulcers, especially in growing pigs or lactating sows,[7] though this is rare.

Many brands are available, including diets specifically formulated with smallholders in mind. When choosing a diet, ensure that you are happy with the ethics of the feed. Look at including fish meal, imported soya, processed animal protein and genetically modified (GM) ingredients after doing some research on each. To ease transition when changing diets, change gradually over a few days by mixing the diet before switching completely.

Own Diet: Some smallholders mill and mix their own diet, but they must work with a nutritionist to ensure that it is appropriately balanced.

Diet Supplementation

Some smallholders feed a commercial diet primarily, but supplement the diet with fruit or vegetables. This must be done carefully, as even small supplementations will disrupt the balance of nutrients being fed, or can lead to too much or too little food being provided. If supplementing the diet, be sure to make sure that nothing that is fed to pigs has passed through a human kitchen.[4]

Other smallholders will choose a more formal dietary supplementation for one feed per day, for example using a cereal such as wheat and a protein source such as beans and peas. If you have a plentiful supply of cereal this may work well for you, but please consider the points on careful formulation of commercial diets, and whether it is wise to disturb this balance. The advice of a professional pig nutritionist should be sought to determine what may suit your holding and your pigs. There are caveats to most 'cheats' – for example, protein sources such as peas and beans cause digestive upset if fed in quantities that are too high.[6] Commercial feeds that include pulses such as beans (instead of importing soya) will cook and dehull them to reduce their anti-nutritional effects.[12] For lactating sows, any supplementary feed must be decreased and a commercially produced diet increased to ensure that they can cope with the demands of lactation.

Straw or sugarbeet can be used to supplement a pig that is on a very small volume of compound food – for example a pet pig or a dry sow – to help them to feel full even though they are not eating very much.

Glen Holloway at the Mill House Cider Museum in Dorset supplements his pigs' diet with the apple pulp left over from cider making in the autumn. This pulp has not, of course, been anywhere near a kitchen.

FEEDING FOR GROWTH

Pre-Weaning:

Milk is the only nutrient source required for very young pigs, and is best provided by the sow. Piglets require lots of energy (delivered through milk) due to their incredible growth rates: they more than double their birth weight in the first week, and often triple it by the end of the second week.[13]

These four-day-old piglets are having a 'power nap' after a good feed.

Struggling to keep up with these demands, the UK pig sector loses about 12 per cent of piglets before weaning.[14] Many piglets are lain on by the sow, often too weak to move out of the way as they haven't drunk enough milk. Many other piglet deaths are due to starvation and hypothermia, both as a result of inadequate milk intake. The steps required to try to ensure good colostrum and milk intake of piglets, as well as how to supplement this, are covered in Chapter 13, 'Breeding'.

Piglets very easily become iron deficient and develop anaemia. The condition is covered in Chapter 9, 'Diseases of Pigs', but it is important to note the nutritional component of this disease. Iron is found in soil, and many indoor breeders will provide piglets with earth for them to root in. Others will supplement with injections of iron (*see* Chapter 13, 'Breeding').

Research has examined how to ease the transition from milk to solid feed for piglets, and has found that feeding suckling piglets a 'creep feed' (high protein, very tasty and often milky piglet diet) is hugely beneficial to gut health later on.[15] This is really important on smallholdings where piglets are weaned much later than commercial holdings, so must be eating solid food before they are weaned as the sow's milk production will start to decrease.

CREEP FEEDING

Creep feed doesn't cause piglets to drink less milk. In fact, piglets that eat a lot of creep also drink a lot of milk.[13] If you have a sow losing condition unacceptably over her lactation, creep feeding won't 'take the pressure' off her: her feed intake needs to be increased, or the piglets need to be weaned.

A starter pellet can be offered from about one week of age. Outside it can be offered in piglet feeders that sows cannot access,[16] or in a creep area inside or outside but away from a heat lamp. Creep must be refreshed daily as piglets particularly like this fresh, and it must not be left to attract vermin. Mixing it with water into a 'porridge' can double the feed intake but it must be replaced twice a day. Sue and Stephen find that piglets approach about 1lb a day by weaning. Creep feeding is even more important when the sow's milk production may be lower, such as in the summer.[2]

Post Weaning

Commercially reared pigs will be fed six or more different diets from weaning to slaughter. Each is intended to provide the pig with its ideal nutritional balance, to enable the best possible feed conversion ratio. The feed conversion ratio describes the volume of food necessary to put on one kilogram of meat.

After several months of growth this laying down of muscle will slow in preference to laying down fat to add flavour to the meat. Commercially the pig can be slaughtered when this lean muscle-to-fat ratio is at the perfect balance.

In contrast, most smallholders do not intend to achieve the most efficient growth or the leanest carcass. In fact, some argue that slower growth leads to better fat marbling and therefore better flavour, so many choose to feed sow food to growing pigs (about 15 per cent crude protein). This doesn't tend to cause any more problems than a reduced feed-conversion efficiency. Sow food isn't very different in composition to a finisher pig diet, so if you want to achieve slightly better efficiency, then feed a more tailored diet to younger pigs and then swap to sow food for finishing.

Commercial pigs will be fed ad lib and will not go fat due to the careful nutritional balance of their different diets and their genetic make-up for maximal growth. Smallholders will rarely feed ad lib as pure breeds will reach full size earlier and tend to go fat more easily. In terms of food amount, an ideal weight per day of feed is impossible to say, and you should refer to guidelines on the food label and receive recommendations from others with your breed or type of pig. 1-2 per cent of body weight is a good starting point, or many breeders feed 1lb (just under 0.5kg) per month of age until five months of age, then keep the amount constant until pigs go to slaughter. Adjust this based on the body-condition score of your live pigs and the back fat of any slaughtered pigs. Trial and error will be necessary over a few batches, so please don't expect to achieve the perfect carcass balance first time.

VARIABLES THAT AFFECT BODY CONDITION

Different producers will have different vegetation covers on their paddocks, have breeds of different sizes, have different sized paddocks for activity levels, be operating at different temperatures, and be using feeds with different nutritional value.

The same feed should be given for at least one week after weaning, then pigs should be gradually moved on to their new feed. Weaning is a really stressful time for a piglet. Many will pause eating for a few days and their growth will take a knock. The later the weaning, the lower the stress for piglets, but this must be balanced with sow health.

FEEDING SOWS FOR REPRODUCTION

Sows that are too fat are more likely to experience farrowing problems, and are less likely to eat enough during lactation, which results in weaker litters at weaning.[17] However, due to the huge amount of energy required to feed piglets, sows will drop some condition over a lactation, so must not be too skinny when they farrow. This weight loss can be even more pronounced on smallholdings where sows feed piglets for a longer period. Sows that are too thin will then struggle to come on heat and maintain the next pregnancy.[17] Weight loss is often more severe during the summer, when warm weather leads to lower feed intake.

Sows should farrow at a body condition score of 3 to 3.5.[17] We do not want to reduce feed intake when a sow is putting energy into growing piglets, and therefore the sow should be at a body condition score of 3 by the end of the first trimester, at around week five of gestation.[17] They should finish lactation at a body condition score of 3 to 2.5.[17]

To achieve this through feeding, sows not feeding piglets should be fed 2.3 to 3.5kg per day. This will vary for the season, and more energy will be required for temperatures below 20°C.[11] Edwards[6] recommends 2 to 2.5kg per day in the summer and 3 to 3.5kg in the winter. This is known as your 'base level'. Sows do not have vastly different energy requirements until the last month

This sow is feeding nine six-week piglets and has kept her condition well.

of pregnancy, when the most piglet growth and udder development occurs.[6] At this time the base level should be increased by 0.5kg, unless the sow is already fat. If sows suffer with constipation at farrowing, you can replace 0.5kg of the diet on the day of farrowing and the day after farrowing with oats or bran to help with this.[2]

After farrowing, milk production will increase to peak lactation (the most milk the sow will produce) several weeks post farrowing and then gradually decline. To meet this huge energy demand, feeding should be gradually increased by 0.5kg per day from farrowing to the maximum that she will eat.[6] Expect sows to eat approximately

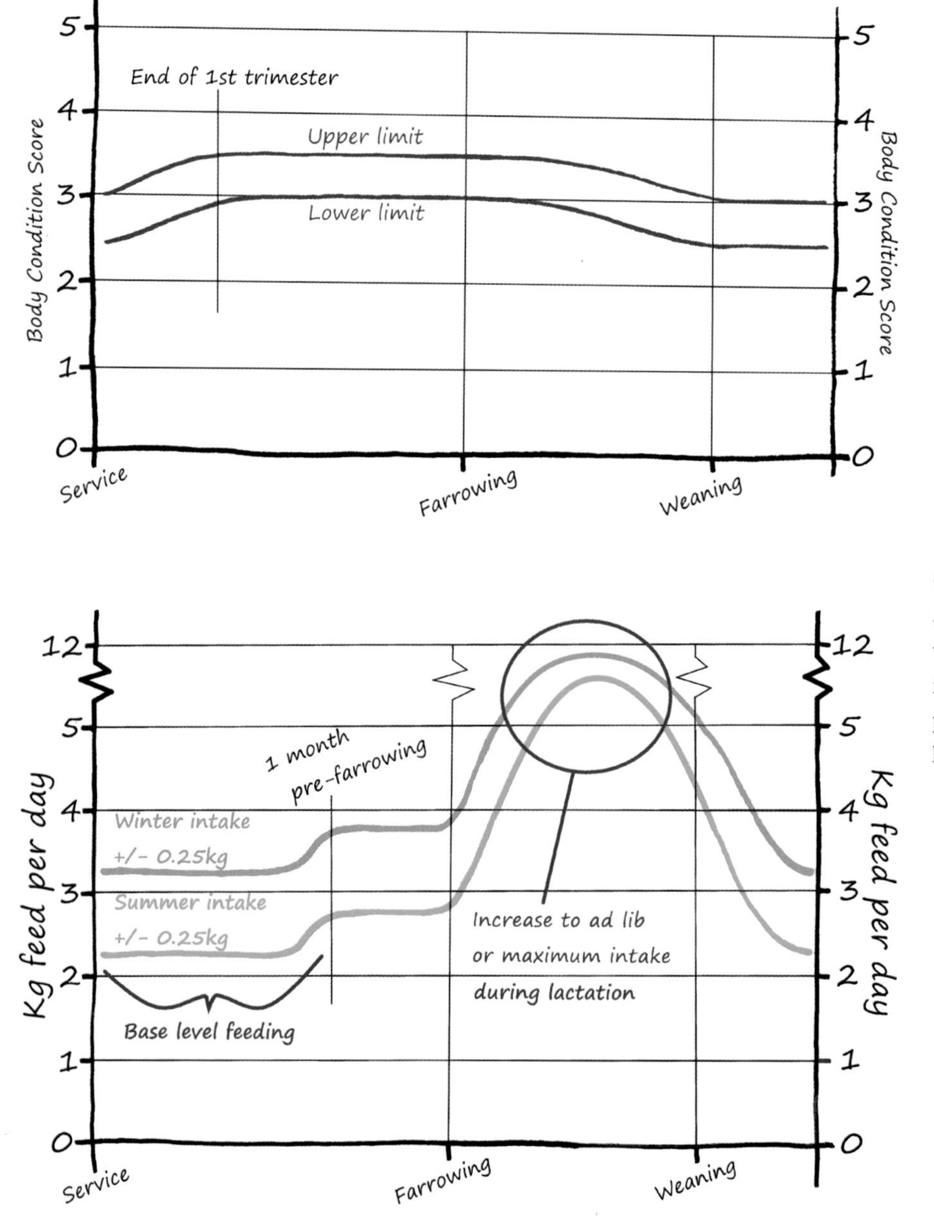

The acceptable range of body condition through the sow's reproductive stages.

Graph showing an overview of the sow's feed intakes during pregnancy and lactation.

an extra kilo of food above the maintenance ration for every piglet (usually up to 12kg per day).[18] Sue and Stephen achieve around 10kg feed and 3.5 to 4kg rolled barley. They add the barley as it is very palatable and it encourages the piglets to start eating, and encourages mum to eat enough as well.

Ad lib feed is ideal but should be balanced with the risk of vermin. Instead, feed can be split into as many feeds per day as you can, at least twice a day but more in the summer.[15] In very hot weather, feed at the coolest times of the day, and try wetting the feed to see if the sow will eat more.[19] The diet should be high quality (preferably specific to lactation), with very little roughage or supplementation. Clean, fresh and plentiful water is also crucial to meet these milk demands.

Post-weaning, reduce feed gradually. If you are breeding the sow straightaway, keep the amount of feed fairly high (more than 3kg per day) up to mating to increase litter size: this process is called 'flushing'.[6] Then take this down to your base level and continue to feed at this rate. Flushing can also be used around first service in gilts with an aim to increase litter sizes, by increasing feed intake for the fourteen days before service.[20] The numbers provided are for larger breed, rare sows. Smaller breeds may require less food.

One of the beauties of smallholdings is that maximal efficiency is usually not sought after. Therefore if a sow comes out of lactation at an incorrect body-condition score, simply get her to the appropriate score before breeding her again.

FEEDING: MORE THAN JUST NUTRITION

On top of providing nutrition, feeding can provide enrichment and the freedom to express natural behaviour. Pigs love to forage, so ensure that they have grass (or hay or haylage if the grass is gone, unless they

These pigs are enjoying a horse 'jolly ball' for some extra enrichment.

are lactating). Resowing paddocks with forage crops, such as legumes, can really encourage this. Sue and Stephen find the Cotswold Seeds Pig Rooting Mix really helpful.[21] Ensuring that your paddocks are rotated, rested and resowed with legumes will also increase the amount of nitrogen that can be fixed in your soil, reducing the harmful environmental impact of your holding.[22]

Enrichment can also be provided with feed balls, or throwing in other items such as cardboard boxes for them to destroy. Always ensure that any enrichment provided is safe for pigs to eat if it could be destroyed, as it will often come to that.

FEEDING THE PET PIG

Pet pigs are very often overweight, to the point where their health is negatively affected. An overweight pig is likely to develop arthritic joint disease later in life. This is especially the case where conformation (breeding) is poor, meaning that their weight is not balanced naturally across their limbs. They are also likely to develop skin issues in their folds of skin, they may be unable to see or hear due to fat around their face, and they may become cast (stuck on their back) and not able to get up.

In addition, they are more likely to struggle in warm weather.

Pet pigs are generally best fed a specific pet pellet, designed to include plenty of fibre so that more bulk can be fed whilst being less nutritionally rich. This means that the pig can feel full, but is less likely to become obese. Pet pigs should be fed small amounts depending on their breed, activity level and grazing, but with 1–2 per cent body weight as a rough guide. Many pet pigs will live off mainly grass and should be carefully rationed. If you are giving treats to pigs, be sure to adjust their main feed accordingly.

Body-condition scoring must still be used in pet pigs, but a few more areas can be examined.[23] Pigs' eyes should be clearly visible, not sunken or hidden in skin folds.

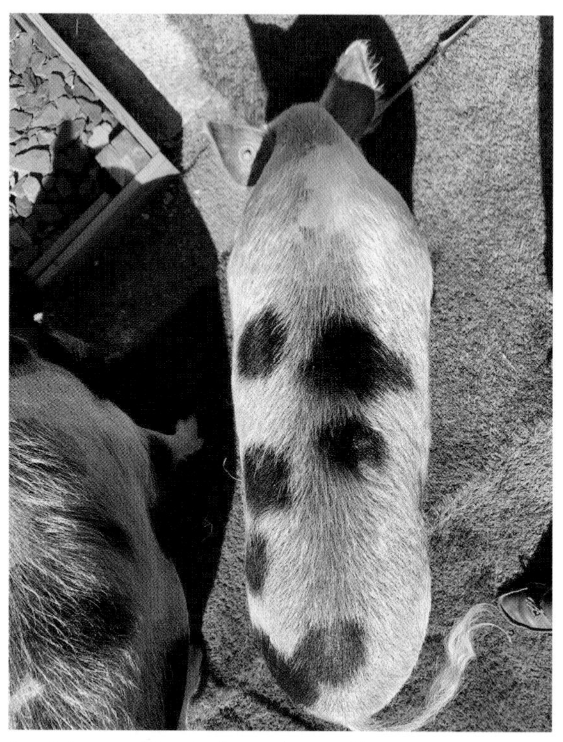

Looking at this pig from above allows us to see that he has a waist, and his shoulders are not too much wider than his head. Therefore he is a good weight.

This pig is overweight because his eyes are hidden in skin folds, his cheeks are bulging, he has folds of skin over his forehead, and rolls of neck fat.

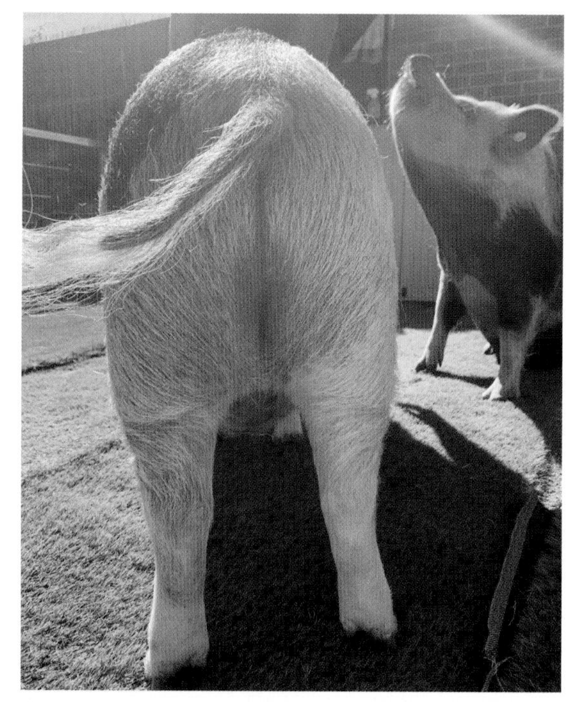

A pet pig can be condition scored from this angle.

This pig has a clear neck and his belly is not too pendulous. He is a good weight. He is also experiencing a totally normal Kunekune moult.

Ears should be upright in most breeds, not pushed forwards into a more horizontal position by fat. Cheeks should not bulge, and pigs should not have folds of skin over their forehead. Pigs should have a true neck, not rolls of neck fat. Their shoulders should not have a fatty hump, and their ribs should be felt (but not seen) easily under the skin. In general, their belly should not hang below the knees. Even in pot-bellied or pregnant pigs it should not touch the ground. You should be able to see that the pig has a waist from above and be able to see the vertebrae over the shoulders. Lastly, the tail base should sit at the same level as the rump and not be buried in fat.

Exercise and slow feeding should be encouraged using food-dispensing toys or scatter feeding, whilst keeping vermin prevention in mind. Pet pig enrichment activities include filling a sandpit or cardboard box with straw and hiding pig nuts amongst the straw. You can also use feed to do training with your pig - for example, clicker training your pig to tolerate procedures such as nail clipping, or learning some fun tricks.

CHAPTER 6

HANDLING

Pigs can be dangerous. They are heavy, easily spooked, and can run really quickly. They can give painful bites, and boar tusks can cause serious injury. Therefore handling must be done with safety to the fore at all times. Handling must also be such that a pig feels as little stress as possible: remember that it is in the Code of Practice[1] that 'excessive force must not be used', which includes striking or kicking pigs for any reason. Poor pig handling really does waste the day, and investing time in making pigs amenable will help you in the long run. Smallholders should have time to do this, and they should seek to understand the best way of moving and handling each of their pig personalities. For example, getting your pigs very used to you being close to them whilst eating will mean that, if you need to inject or slap mark, you may be able to do this with just a bowl of food.

The best way to lower pig stress is to get pigs used to human interaction.

Much of our pig-handling knowledge has come from Temple Grandin's ground-breaking work, which has led to happier pigs globally.[2] Like much science, some has later been disproven, notably any suggestion of exerting 'dominance' over pigs, either verbally or physically. Alpha or dominance theories were incorrectly applied to dogs after a very small study on wolves in the 1940s[3]. We now know that dogs are not like wolves, and pigs are definitely not like wolves! Pig social hierarchies are incredibly complicated and not something we will explore here, so instead, keepers are encouraged to use positive reinforcement-based training methods to build good relationships. For more on this, go to https://www.positivelypigs.com/.

'HANDS-ON' PIG HANDLING

AHDB's online course, 'Moving and handling pigs',[4] has informed much of this chapter, but written words will never be as valuable as going and doing it yourself. Don't let the commercial nature of this course put you off. The content is transferable to all settings, including pet pigs. Find out more and complete the course here: https://ahdb.org.uk/moving-and-handling-pigs.

THE SCIENCE BEHIND PIG HANDLING[4]

Pigs are a prey species: Pigs are considered a prey species, meaning that they are caught and eaten by predators. Therefore they are most likely to run from danger. This does not mean that they can't be aggressive! Even the most amenable sows can be very protective of piglets, and boars should always be treated with suspicion.

A pig's visual field: Because a pig's eyes are on the side of its face, it can see right round

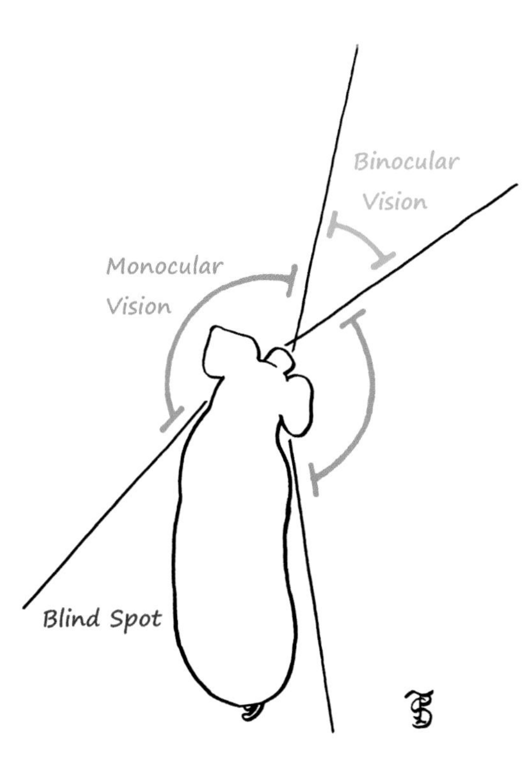

The pig's field of vision.

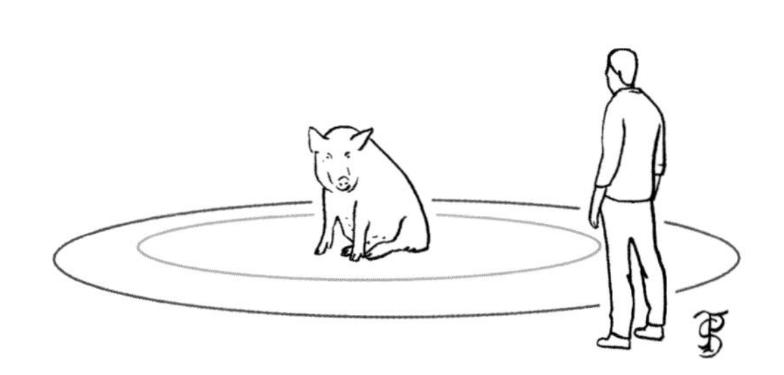

A stressed or unsocialised pig will have a larger flight zone than a happy and relaxed pig.

to its rump, though there is a blind spot directly behind it which should be avoided. Pigs see well straight in front of them, but don't have the concentration of both eyes as we do, so can struggle with depth perception. This means that they may need time to 'work things out', such as floor surfaces, unexpected objects or trailer ramps.

Response to movement: Pigs will see sudden movement as a greater threat than a smooth and predictable one.

Use of hearing and smell: A pig's ears will point and turn towards a perceived threat. If the threat is in front, the ears will be forwards; if behind, the ears will point backwards. Pigs frequently stop and put their nose to a new area, to smell it or look at it closely. Loud noises will startle pigs.

Flight zone: 'Flight zone' describes the area around a pig that it will actively keep free from a perceived threat. In the presence of a threat you will see the pig manoeuvre itself to keep the threat out of this zone. This zone varies considerably between pigs: for example a stressed pig will have a larger flight zone. In a small area this can lead to stressed pigs being forced to bunch together away from a threat, or to circle around the threat to escape it. We should keep out of this flight zone wherever possible to relieve stress on the pig.

Pigs are a herd species: Pigs, and especially piglets or scared animals, do not like to be separated from the herd. Piglets should be moved as a group to avoid stress. Pigs like to move with another pig next to them.

RECOGNISING A STRESSED PIG[4]

Relaxed pigs will walk quietly with the herd, with a low head carriage and ears down. Stressed pigs will have their ears pricked and the head carriage will be high. They might circle, bunch and pile on top of each other. They may vocalise with a high-pitched squeal. An adult boar may chomp or growl at this point, which indicates that you should move out of his space. This is when accidents and injuries are likely to occur as the situation will no longer be controlled.

TOP TIPS FOR GOOD PIG HANDLING[4]

Make Yourself Unthreatening

Pigs can be unpredictable in terms of what they perceive as a threat. However, they will be able to tackle objects or changes in area if they are given time to investigate, and as long as there is only one threat at a time. If you are seen as a threat, and are also moving them through a gate from behind that is also seen as a threat, this could well become too much for the pig and cause it to run to the side to escape both things, rather than move through the gate. If you are not seen as a threat you can use a bucket of feed to move it through the gate. This can only be achieved by taking time to handle your pigs, ensuring that there are more positive interactions between you than negative ones.

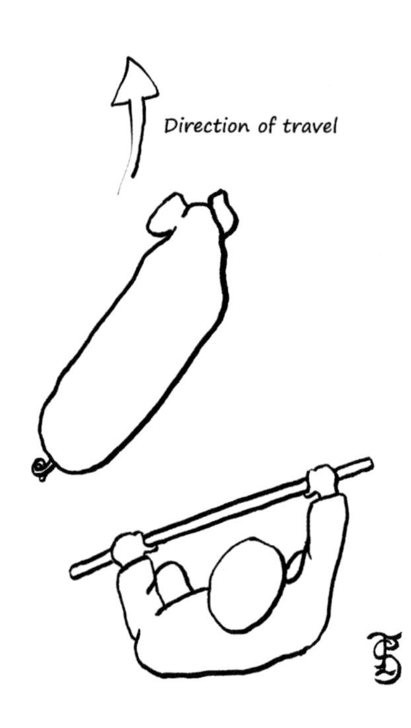

Standing behind a pig's shoulder will cause it to move forwards.

Use the Science

Use the science to your advantage. For example, holding a sorting board so that you are more visible and standing at the edge of a pig's flight zone behind its shoulder (but not in a blind spot) will cause it to move forwards. Standing in front of its shoulder will cause it to move backwards. Standing in a pig's blind spot will make it nervous as it will know that someone is there, through its other senses, but won't be able to see them. Standing close to boundaries, such as the edge of gate posts, will allow pigs to have the space they need to move around you, and they will therefore be more likely to move through the gate.

Here, Stephen is taking time to socialise a group of weaners. He is sitting so they feel comfortable with his presence, and is letting them approach him.

Use Forward Planning

When designing a holding, consider how you will restrain/move/load these pigs. Also consider possible social issues between pigs. For example, think very carefully before putting yourself between fighting pigs by considering your own safety, balanced with pig safety. By this point it may well be too late.

The next stage of planning takes place just before the event. Consider the following:

The Pigs you are Moving

How amenable are the pigs? Will they follow with food or will they need to be guided

At around 400kg, Bran the boar needs a consistent and very careful approach.

from behind? A sow that is protective of her young, or a boar, will need different strategies and extra precautions compared to a couple of weaners, such as prior considerations of escape routes, use of sorting boards and protective footwear. Boars require a consistently wary approach and an experienced pig handler. If you are hiring a boar, speak to the owner and learn how they handle him. Imitate that approach wherever possible to avoid incidents.

The Route you are Taking

Where will you need guides for pigs to ensure they are going the right way? What are the surfaces like? Where might pigs need to stop and investigate? Is the route clear of obstacles? What is their final destination? Is it familiar, or are they likely to become stressed? By considering these stresses in advance you might decide that you need to prepare for the event months in advance. For example, to prevent later stress during loading, in advance and over a few sessions, feed the pigs in the trailer so they can get used to it gradually without any scary consequences.

Use Patience

Movement around pigs should be predictable, quiet, slow and without force. Don't use flappy or shiny objects, such as feed bags or plastic piping. Don't yell or push pigs. If things aren't working, take a moment to pause and think. This will give the pig time to calm down, and allow you to evaluate your options. If a pig pauses, give it time to do so, without prodding it continuously. A light and brief pressure of a hand on their back lets them know that you want them to keep going. They will then move off once they have investigated where necessary. Before encouraging a pig to move, always check that their path is clear.

The use of electric goads on farms is almost always illegal[1] and should not be considered.

Use Aids

Pig sorting boards: These should never be used to hit a pig or to push it along. They should be held still and moved slowly so that the pig isn't pressured. They are an aid to be used alongside an understanding of pig behaviour (especially the visual field) and all other tips.

Pig boards are useful for blocking gaps as they are a far better visual cue for a pig than just a person. They should be held touching the ground so that a pig can't get its nose underneath. A pig board can be used to block a pig from backing up by slowly moving this behind the pig. Stand still and wait for the pig to assess the situation. If done without pressure, the pig will see the bright object and decide that a forward path is most appropriate. This video (https://www.youtube.com/watch?v=uVPTKzEafKc, Sorting board training by the Pig Placement Network) shows excellent use of a sorting board in a smallholding /pet pig /sanctuary setting. Note how the handler stands away from the pig to the side, at around the level of its flank, to get it to move forwards.

Obviously, if put under a great deal of pressure a pig will charge at a person with a pig board, and if strong enough, will push the person out of the way. Therefore, pig

For a few lessons on the successful use of a pig board in very difficult circumstances, go and watch the pig classes at a livestock show!

boards are best used to guide an animal into a proper restraining area, rather than as an implement of restraint in itself.

Sorting paddles: Paddle boards are little plastic paddles on a pole with grains inside which will make a quiet noise if shaken. They are most useful to move groups of pigs through passageways, which won't be relevant on most smallholdings. They can be used alongside a sorting board to give the pig a clearer indication of where you want it to move, and those intrigued should watch the relevant section of AHDB's online course, 'Moving and handling pigs'.[4] It must be noted that they should never be used to hit a pig or cause fear in an animal.

Use Humanity

Pigs should never be struck with a body part or object. This includes when trying to get a sick pig to move. It is inappropriate to get a pig up by its tail or to restrain or move a pig by holding its ear. If you have a situation with a pig that you do not know how to rectify, speak to your vet, as vets receive a lot of training on animal handling.

RESTRAINING PIGS

Pigs may need to be restrained for a variety of reasons, such as an emergency vet visit to be examined, or to be given an injection. The minimum restraint necessary to complete a task is best in order to minimise stress for both pigs and handlers. Ear defenders may be necessary when restraining pigs.

Before restraining a pig, the pig, the task and the space available for handling must be considered.

The pig: Every effort should be made to train pigs for cooperative care, which will reduce the restraint required. A sick pig may require less restraint, but a pig in pain may require more skilled handling. If a pig needs to be moved but cannot stand, speak to your vet about the best course of action.

The task: The task will dictate the restraint needed. Taking a respiratory rate in a pig requires no restraint (restraint is actually detrimental), whereas the use of a captive bolt stunner requires a pig to be very still and therefore more restraint will be needed. If the task might be a painful procedure, or one that the pig is unlikely to be amenable to, it may be necessary to contact your vet who may restrain it with a sedative. Ensure that the task has been properly planned so that the pig will remain restrained for the minimum amount of time possible.

The space: Where is the pig currently situated, and will restraint be possible there? A suitable space for restraint is small enough that the pig's movement is restricted, but large enough that the activity can take place safely. There should be good lighting and a non-slip floor. A trailer with a side door can be useful, with the pigs pre-loaded after being trailer trained well in advance. An arc is not a suitable area to restrain or examine a pig: the confined space increases the likelihood of getting trapped in the back of the arc with an angry pig, and the lighting is poor.

Methods of Restraint[5]

The methods described below can be used to restrain a pig.

By Holding

Piglets can be picked up and restrained for most procedures, excluding euthanasia with a captive bolt stunner or firearm. Piglets can be picked up and held close to your body. Try to pick them up with two hands supporting their chest and torso. Catching a back leg can aid in this in piglets less than 10kg, but the chest should be immediately supported with the other hand without any swinging. After, piglets should be placed gently back on to the floor.

The correct way to hold a piglet.

Using a Feed Bucket

Especially for pigs not fed ad lib, a bucket of feed can work wonders for tasks such as injections or slap marking.

Using a Belly Rub

Many pigs will lie down for a belly rub, especially pet pigs, and should be trained to do so wherever possible. A pig that loves belly rubs can be distracted from procedures such as a physical examination or a foot trim.

A belly rub left this pig immobilised for quite some time.

Using a Small Space

A small space can be useful when it is necessary to hold a pig. Provided the pig is calm and well handled, a well-positioned gate and hurdles, or in some cases a pig board

This makeshift handling system was adequate to allow Pete to detusk this boar without sedation or a snare.

Twiggy being calmly restrained to receive an injection.

Trailer training pigs can give another option for pig restraint, such as to give an injection.

for smaller pigs, can achieve this. However, if the pig is stressed it will likely break free, possibly causing injury.

Using a Snare

A pig snare is a device that loops round the upper jaw of the pig. The pig's natural reaction to this is to sit back, and therefore the pig can be restrained. Snares should only be used where less invasive methods of restraint are not suitable for the pig or the task. Only use a device that is specifically meant for this purpose.

Snares come in both short-handled or long-handled versions. Many long-handled snares are 'lockable', meaning that you have to press a button to release it. Short-handled snares tend to rely on the lever arm that you create to keep tension on the snare.

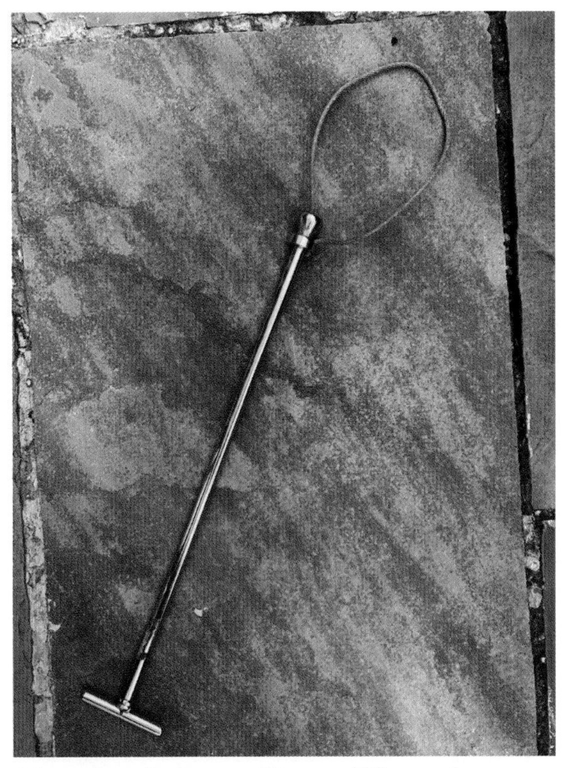

A long-handled snare. Many will have a button that releases the snare. Be sure to explain this to the person you are passing it to so that they don't accidentally press it.

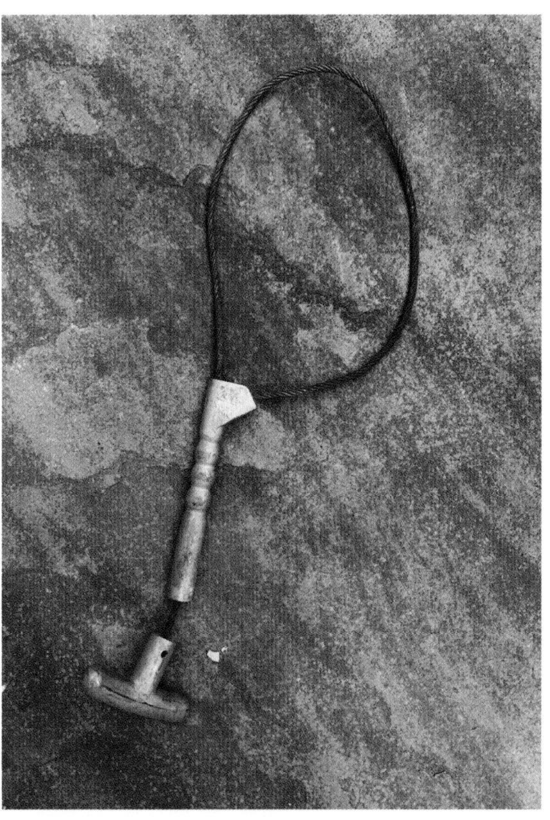

A short-handled snare is easy to clean and disinfect without causing rust, and is less of an accidental weapon should it not be released correctly.

USING A SNARE

1. Start with a cleaned and disinfected snare that has been checked for serviceability. Ensure that there are no sharp edges or fraying. Ensure that you understand how to tighten and loosen the snare design that you have.
2. Prepare the space and any equipment necessary for the task. Operating in a small space will ease the challenge of catching the pig, but you must have enough space to complete the task. A stressed pig will be less interested in a snare, so try not to stress it out whilst moving it into the space. Start with your

pig just slightly in front of a wall, which you will use to your advantage shortly.
3. Have a spare person to hold the snare while you complete the task. Depending on the size of the pig, this may require strength, so ensure that they are fit for the job. For very big pigs, it can help for the person to have a post or similar to hold on to. Ensure everyone present is wearing ear defence, and warn all parties that the pig may make a very loud noise.
4. Stand in front of, but to the side of the pig, with the snare loop a little bigger than the jaw it needs to go round.
5. Wait patiently for the pig to be inquisitive about the snare, and eventually it will put the loop into its mouth. Smearing some banana (not from a human kitchen) on to the loop can help.
6. Push the loop further into the pig's mouth by walking towards it. Here the wall behind it will stop it from endlessly walking backwards with you. Wait until you have the snare fully behind the upper canines.
7. Tighten the snare and step back, holding the snare handle straight ahead of the pig at its head height. The pig will pull back against you, making a lever arm. Pass the snare handle to the helper, and complete the required task as quickly as possible. Never tie the snare to anything or fix it to a point. Never move a pig using a snare.
8. Release the snare by smoothly easing the mechanism.

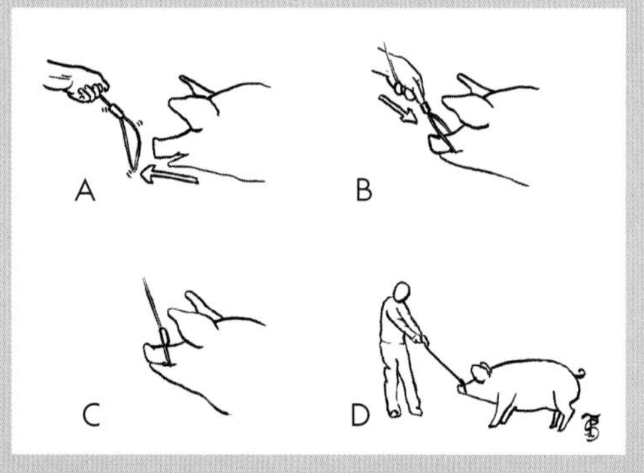

The steps necessary when applying a snare. Note that the snare is moved right to the back of the mouth before tightening.

A snare won't work for very small piglets (less than about 10kg) as it will just flick off their snout. You can fashion the same mechanism with bailer twine, but we only use this method for captive bolt stunning where we need the head perfectly still and away from any human body parts. The rest of the time it is generally most appropriate to pick up the piglet instead.

Enticing a pig to engage with the snare.

Tightening the snare.

The snare is now applied.

This pig can now be held comfortably.

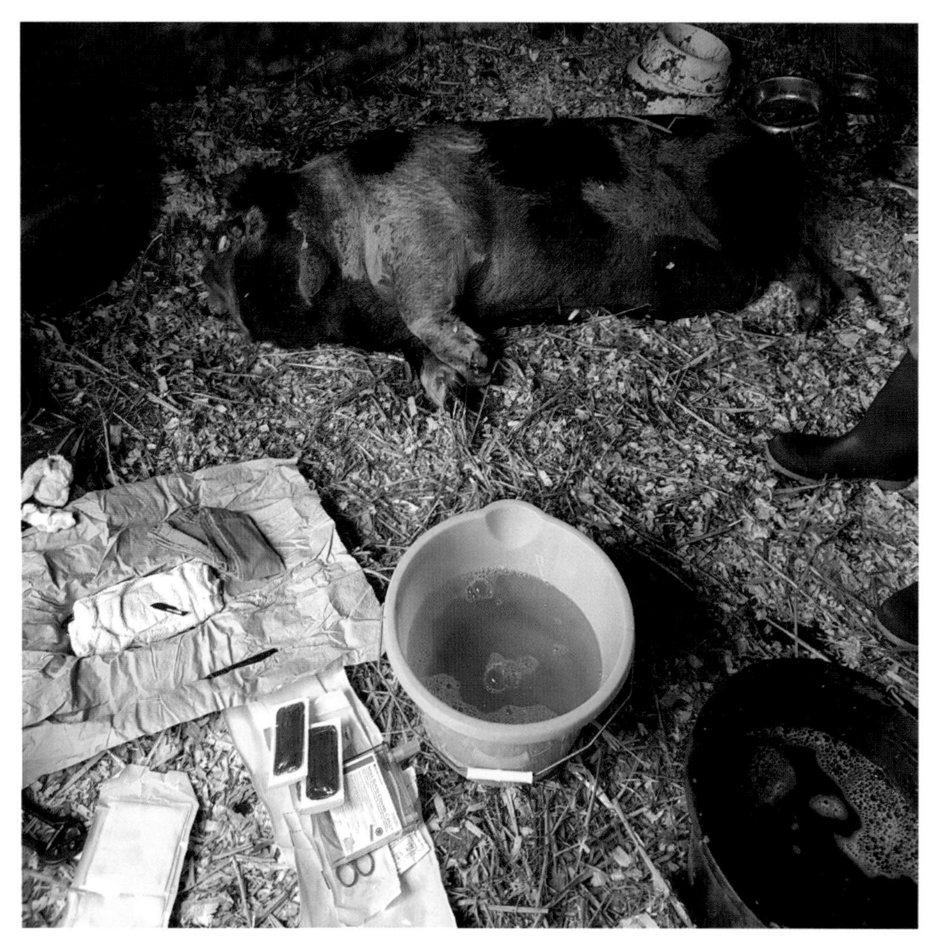

Despite this pig being so amenable that his owner could trim his nails with just belly rubs, the painful procedure of removing a lump required a general anaesthetic.

Using Sedation or General Anaesthetic

Vets are very lucky to have medication on their side, which can be used to great effect! Sedation or general anaesthetic is necessary for painful procedures that would be unacceptable to complete whilst the pig is conscious, or for pigs that are not amenable to a task being completed whilst they are conscious. Sedation is just a lighter form of a general anaesthetic, so for some procedures your vet may try just sedation initially, but will increase the dose to a deeper sedation or general anaesthetic as required.

Although this sounds like a golden bullet, please do remember that pig sedation only comes in the form of an injection, so your vet will still need to get close enough to inject. A stressed pig will also need a much higher dose as they will 'fight' the medication. Therefore sedation does not replace the need to train a pig to be amenable to the vet. Furthermore, sedation is not without risk, which should be weighed up alongside your vet.

Your vet will give full instructions before a visit where sedation may be required, but you will need to starve your pig beforehand to reduce the chance of it vomiting and accidentally breathing in the vomit. Your vet will remove the pig's water just as it starts to become sleepy, and this shouldn't be returned until it has fully regained consciousness. The medicines that we use also dilate blood vessels and can cause a pig to get very cold, especially piglets. Therefore your vet may advise special measures to keep pigs warm.

CHAPTER 7

TRANSPORTING PIGS

Before transporting a pig you need to consult the relevant legislation, as well as taking into account the needs of the pig. We will not cover import and export rules here, but will cover the transport of pigs within the UK from a smallholding context - that is, journeys under twelve hours, and not including journeys in transporter lorries.

THE WELFARE OF ANIMALS (TRANSPORT) (ENGLAND) ORDER 2006

The Welfare of Animals (Transport) (England) Order 2006[1] states that 'no person shall transport animals or cause animals to be transported in a way likely to cause injury or undue suffering'.

ADMINISTRATIVE LEGISLATIVE REQUIREMENTS

All journeys require a movement licence, an animal transport certificate, a standstill period and identification during transport.

Movement Licences[1, 2]

If a pig is moved it must be under licence, and this licence should be in place before the movement takes place. The keeper must obtain the licence, and once the pigs arrive at their destination, the new keeper should confirm the number of pigs received. This is done through electronic systems in England and Wales (eAML2) and Scotland (ScotEID), or paper records from DAERA in Northern Ireland. Pigs in transit must be accompanied by the movement documents, which will be emailed to you before the movement takes place and should be printed.

Setting up a Movement via eAML2

In order to register with eAML2 you will also need to register with Pig Hub. This holds all your information, such as the details of your vet, and uses this information to fill out the eAML2. You can then register with eAML2. The eAML2 website can be found at https://www.eaml2.org.uk/ami/home.eb and the Pig Hub website can be found at https://www.pighub.org.uk/iip/home.eb

MOVEMENT LICENCE EXEMPTION

The law states that 'a keeper moving a pig to a veterinary practice for emergency treatment need not notify the movement'. Note that this exemption does not apply for non-emergency veterinary treatment.[1]

Animal Transport Certificate[3]

For all journeys, an animal transport certificate is required, which should record the following:

- origin and ownership of the pigs
- place of departure and destination
- date and time of departure
- expected duration of the journey

You can contact APHA's welfare in transport team[3] for a template.

Standstill Period and Identification During Transport

The legislation described in Chapter 2, 'Legislation and regulation around UK pig keeping', applies.

Non-Economic Journeys or Economic Journeys under 65km

An economic activity means one that is part of a business or trade. Therefore, if you plan to sell pork then your pigs are part of an economic activity. There are extra stipulations for economic journeys over 65km (just over 40 miles), so it is sensible when starting out as a producer to choose a breeder and an abattoir that are closer than this.

For these shorter or non-economic journeys, all that is required is a movement licence to be in place, as previously described, and an animal transport certificate. That being said, courses that allow transporters to achieve a certificate of competence necessary for longer journeys cover topics relevant for anyone transporting a pig, and therefore completion is encouraged.

Economic Journeys over 65km

For these longer economic journeys a Type 1 Transporter Authorisation will be needed for journeys over 65km and up to eight hours, and a Type 2 Transporter Authorisation for journeys over eight hours. A transporter authorisation is free of charge, and an application pack can be obtained from APHA at WIT@apha.gov.uk.

To achieve this you will also need to apply for a Certificate of Competence, which is generally a theory test for journeys under eight hours and a practical assessment for journeys over eight hours. Details of this can be found at https://www.gov.uk/guidance/animal-welfare#certificate-of-competence

The training and examinations for the certificate can all be done online for around £200. There is one main exam, then supplementary exams for each species: pigs, cattle, sheep, goats and horses. Some institutions include all species in the price, some charge extra. Once achieved, you must then put this knowledge into practice.

If you are using a transporter, it is your responsibility to check they are authorised. More information can be found here on how to do this: https://www.gov.uk/guidance/animal-welfare#transporter-authorisations.

PRACTICAL TIPS FOR TRANSPORTING PIGS[4]

Good pig transport involves looking at pigs and understanding their individual requirements. Good relationships with your pigs will assist your decision making.

MINIMISE STRESS

Minimising stress for pigs is especially important during transport. Some well-muscled rare breeds, particularly the Landrace and Pietrain, can suffer from a condition known as malignant hyperthermia or porcine stress syndrome, in which they have a catastrophic cellular reaction when they become stressed. This has led to many pigs not making it to the end of their journey. Stress before slaughter also leads to meat becoming pale, soft and exudative (PSE), which decreases meat and carcass quality.

Planning a Journey

Choice of Vehicle

Vehicles must be well maintained and checked between uses, including checking tyre pressures. Transport must be suitable for the specific pigs being transported, and can range from a large dog crate in the boot for a couple of weaners to a trailer for larger pigs. The following factors must be considered:

Thermal comfort: Transport must offer protection from weather, but also have ventilation.

Space: Pigs must be able to comfortably stand up and lie down. They should have enough space to prevent injury. Smallholders are unlikely to be able to get pigs off transport to check them en route, so the transport also needs to be big enough to be able to perform a thorough check of pigs. In hot weather, pigs should be given more space. Some suggested minimum values can be found on p.30 of the European Commission's[4] guide to good practices for the transport of pigs.

Other comfort: Pigs should be transported on non-slippery floors. The transport should be light enough for a pig to understand its surroundings.

Sue and Stephen's boar in a large trailer with a good bed. The jockey door enables him to be checked easily and offered water en route with no danger of him escaping.

Use of Trailers

Ramps should have an incline of 20 degrees or less and, where the angle is more than 10 degrees, cleats or similar must be fitted to the ramp to allow pigs to get traction and prevent them slipping (these should be 8in apart for market pigs).[5] Put straw over the ramp to cover up the shine and stop slipping, which could scare pigs. Trailer training pigs will decrease stress. The trailer should be introduced to pigs as early as possible. At first, scatter feed just on the very edge of the ramp, moving further up the ramp each session. Nothing should happen to the pigs and they should be allowed to investigate the trailer as and when they want to. After several sessions pigs will walk confidently into the trailer. Some people will leave the trailer in the field for the pigs to get used to it, but watch out for them scratching or damaging it.

A suitable livestock trailer for pigs.

These pigs are comfortably loaded with rubber matting over the ramp and fresh straw.

Planning the Details

The journey should be thoroughly planned to be as short as possible, and should include contingency planning for unforeseen events such as traffic. Think about potential risks, suitable actions if those risks occur, and what could be done to mitigate against those risks, such as ensuring that you have excellent breakdown cover. The following questions might be useful:

- Do you need to stop for breaks? Where would be appropriate?
- If a pig has a health issue, where are your nearest vets en route?
- Will you need someone to assist you at any point?
- What exactly will you do on stops? Write a checklist so that you don't forget anything.
- What is the weather forecast? How will this affect your journey? Should you be changing your plans in cases of extreme weather? In cold weather, provide more bedding and adjust ventilation where possible. In hot weather, allow for more space, provide more water, provide more ventilation, and consider travelling earlier or later in the day.
- What is the likely traffic? Schedule the journey to avoid times of high traffic congestion in urban areas.

Provisions

Food: Pigs should not be provided with food during a journey as they will become motion sick. Pigs should be fasted before travelling. Five hours is recommended unless moving to slaughter, in which case ten to twelve hours is appropriate.[4]

Water: Ideally pigs will have access to water, but if this isn't possible they can be given regular access to water during stops, when they will be given adequate time to drink.

Bedding: Bedding material should be provided to soak up fluids and provide comfort and warmth if necessary.

Extra provisions: Make sure you have extra provisions should things go wrong en route.

The Journey Itself

Loading

Every pig must be assessed as fit for transport (*see* later) before being loaded, the intention being that the transport must not cause them serious suffering. Handlers should be sympathetic at all times.

During the Journey

- Drive carefully, smoothly and with slow braking to avoid motion sickness and stress.
- Pigs should be inspected regularly, their condition thoroughly assessed, and note taken of how well they are coping with the journey. Panting animals are likely to be experiencing extreme stress, are too hot or are struggling to breathe due to respiratory disease. Use this information to decide on the best course of action for the pig.
- Don't leave pigs unattended in transport vehicles without appropriate ventilation. Bear in mind that pigs find it even more difficult to regulate their temperature than dogs.

After the Journey

* Unload the pigs as soon as possible and check them over. Give them access to water and food immediately, except when unloading at an abattoir.
* Any vehicle used to transport pigs must be cleaned and disinfected immediately after the journey, as a disease control measure. Dispose of any bedding.
* Records must be kept for at least three years, but five years is best practice.

FITNESS FOR TRANSPORT

European law states that 'no animal shall be transported unless it is fit for the intended journey, and all animals shall be transported in conditions guaranteed not to cause them injury or unnecessary suffering.'[6]

Keepers must make a pre-transport assessment that a pig will not suffer serious welfare problems as a result of the trip. The pig must be able to keep its balance throughout the journey. Also consider whether current conditions are likely to become worse and therefore unacceptable during the course of the journey. The pig must arrive at its location in a state that would still deem it fit for transport.

A pig is considered unfit to travel in the following circumstances:

- If it is unable to move independently without pain or to walk unassisted.
- If it presents a severe open wound, or prolapse.
- If it is a pregnant female for whom 90 per cent or more (102 days) of the expected gestation period has already passed, or a female that has given birth in the previous week.
- If it is a new-born piglet whose navel has not completely healed.
- If it is under three weeks of age (unless the journey is less than 100km).

There are exemptions to these rules: for example, moving a sick pig to a veterinary practice (as long as unnecessary suffering is not being caused).[6]

Further Explanation of the Points Above

* Being unable to move independently without pain or to walk unassisted is interpreted as a pig that is unable to

move or to keep its balance (that is, to remain standing and to walk without help). Therefore, the pig would be likely to lose balance during transport. This would include any animal that is severely lame and only minimally weight-bearing on a limb.[7]

* A pig in circulatory distress or with breathing difficulties can also be covered by this point.[7, 8] Thus a pig that is showing signs of heavy breathing, panting, rapid breathing, or whose nose is going red or blue, is unfit for transport.[7]
* A prolapse is when organs are protruding from the body: rectal, vaginal or uterine prolapses are unfit for transport. A severe open wound is interpreted as making the pig unfit for transport if there is profuse and continuous bleeding, or if the wound could be made worse during transport.[7, 8]
* For hernias, either umbilical or testicular, the situation is less clear. Any pig with a hernia that is more than about 15cm, touching the floor, or with broken skin, should not travel.[8] This is due to the risk of rupture and catastrophic injury during transport.
* Ideally, pregnant females should not travel in the final third of pregnancy.

In borderline cases, consider, possibly alongside your vet, the condition of the pig and whether the complaint is likely to become worse during transport. If moving a pig to slaughter, also consider the likelihood of the carcass being rejected, and always consider the risk of transmitting infectious disease. The common mantra 'if in any doubt, leave the animal out'[8] must be followed. Pigs deemed unfit for transport may instead require veterinary treatment.

CHAPTER 8

MAKING THE MOST OF YOUR VET

We strongly advise registering with a vet as soon as you have pigs to avoid inconvenient delays in treatment. We recommend registering with a farm vet, who will be more familiar with pigs than a vet primarily attending to dogs and cats. Try to look for a farm veterinary practice with a smallholder scheme or club as they will have vets dedicated to dealing with the different demands of serving smallholders. Finally, try to look for a farm vet who has an interest in pigs. Local pig keepers may recommend vets who can cater for your specific set-up.

Vets running smallholder schemes will often hold smallholder meetings, which is an opportunity to learn about important topics or learn practical skills such as administering medicines. The vet may visit the holding to write a veterinary health plan with you - we cover the importance of this later on in the chapter - and they may also have a set price plan for smallholders to gain access to all of this valuable support.

Pete tries to foster a close working relationship with his pig clients.

WORKING WITH YOUR VET

After registering, we encourage working with your vet to prevent disease, rather than having to treat it when it occurs. This will lead to more productive pigs, fewer emergency call-outs and lower medicine costs. That being said, emergencies happen with pigs, and even the best preventative plans can't prevent everything.

It is important that smallholders understand the following:

- How to recognise pig health problems with a basic physical examination.
- The reasons to call a vet immediately.
- How to provide your vet with the relevant information.

- How to administer first aid to a pig before your vet arrives, and how to carry out the treatments and aftercare that they may prescribe.
- How to manage the vet visit and the vet relationship so it is as productive as possible.
- How to put all this information together to reduce the chance of it happening again.

It is important to note that, in order for a vet to prescribe on a holding, they must have visited it and will often need to examine the specific pig that is suffering disease before they can suggest medication. This is not the vet being picky or wanting to get more money. The vets' certification body is very specific about the conditions in which vets are allowed to prescribe, and they must meet this in order to remain practising. Those regulations are in place to protect the pigs in question, to ensure that the correct diagnosis is reached and the best course of action decided.

IDENTIFYING PROBLEMS

Record Keeping

On top of those required by law, records can be kept for many variables, from the number of piglets born per litter to pig growth rates. Records allow us to spot trends, and to act before a situation deteriorates. Accurate records allow vets to rule conditions in or out, which means they will be able to form diagnoses more quickly, more cheaply, and without trialling a whole lot of medicines that may not work.

Records should be kept routinely, as well as when problems arise. They should be recorded in a way that allows you and your vet to scrutinise them with ease. The veterinary health plan visit can be a great time to look into records in more detail, or when any signs of disease arise.

Recognising Pig Health Problems

All pigs must be inspected at least once a day to check that they are in a state of wellbeing.[1] Any time spent looking at pigs is time well spent, but there are times when this minimum will need to be much more frequent (for example around farrowing). For this inspection to be a useful exercise it is important that stock people understand normal pig behaviour and how to respond to changes. It is very likely that you know your pigs and their normal behaviour better than anyone, and this will assist the exercise hugely. Meal times are a great opportunity to check that every pig is eating enthusiastically, as lack of appetite is an early sign that a pig is not feeling well, for many possible reasons.

A change in behaviour does not always mean that disease is present, but it is always important to investigate any such change. For example, it is really helpful to know the timings of a sow's cycle, marking this on a calendar every twenty-one days, as some pigs can act very differently during this time.

If you are worried that a pig may be sick, it will be important to carry out a basic physical examination. This will enable

All pigs must be inspected at least once a day.

A short video or photo can be a great help for a vet trying to make a diagnosis.

you to decide on your next course of action, which will commonly be a phone call to your vet. It will also enable you to give lots of useful information to your vet that will allow them to make a far better assessment of the pig over the phone. Mobile phones make it possible to record physical examinations using photos and videos, which may be a more accurate way of detailing information that can be hard to explain.

The majority of a pig's physical examination is done by looking closely at the pig, without touching it. Pigs can be very easily stressed if they are handled, especially in a way that they are not used to, and this can mask more subtle signs. Parts of the examination that require the pig to do something, such as standing or walking, should only be completed when the pig can do this easily. If any of the descriptions that follow lead to you believing that the pig is in pain, you must pause and call your vet. As you move through the list, write down anything abnormal to report this to your vet.

Pete's cousin Angus can learn a lot about this sick pig just by observing it closely.

Changes in Behaviour

Is the pig responsive to you and showing behaviour you would expect? For example, moving towards or away from you. Is the pig walking in a straight line with good balance and a normal head position? Does it show a normal reaction to a stimulus such as food? Is it interested in food? Is this normal for it? Can it swallow food normally?

If the pig is more sensitive to light than normal, can't stand, or is hiding in the corner of the pen, it's important to phone your vet. To give them more information, describe how the pig is lying. Is it on its tummy or on its side, and possibly fitting or tremoring?

Changes in the Cardiovascular and Respiratory Systems

Is the pig coughing, using more effort to breathe than normal, or is it breathing

As a vet, Pete can use his stethoscope to examine the chest more closely.

with its mouth open? Count the number of breaths per minute. Is the pig pale or do any extremities have a blueish tinge? Does the pig struggle to keep up with others in the group? Are there discharges from the nose or eyes, and if so, of what colour? If there is excessive blood loss, or blood loss that you are unable to stop, always call your vet immediately.

Changes in Locomotion

Does the pig stand squarely with even weight on all four feet? Is it putting less weight on a leg when it walks, or is it reluctant or slow to walk? Can you see any joint swellings or discolorations of the skin in areas that may be consistent with trauma? Do the muscles look even on both sides, or is there muscle wasting? When looking at the pig from the tail end, the spine should be in a straight line and not curved.

It is important to report any very painful lamenesses to your vet immediately - for example a pig vocalising when it tries to stand - due to concerns regarding the notifiable disease foot and mouth. Be aware that pigs can hide the signs of a fracture very well, due to their huge and stabilising bodies of muscle. It is really important to consult your vet if a pig is holding a limb off the ground whilst standing.

Changes in Weight and Shape

Does the pig look thin, scrawny, or a strange shape? Does it have a more swollen abdomen than normal or than others in the group? For adult pigs, routine weight checking (with a weigh tape) will allow you to monitor their normal weight over time. For growing pigs, how do they look in comparison to the rest of the group?

Weight loss indicates that disease has been present for a period of time, either due to reduced food intake or because of a disease that prevents the pig meeting its nutritional needs through feeding. Apart from in our very hairy pig breeds, it is generally much easier to assess weight loss in a pig by eye than it is in a sheep because its body is not concealed by a fleece!

Changes in Defecation and Urination

Look at the faeces in the pen and note the consistency. Look around the hindquarters for any signs of scour or a rectal prolapse. Call your vet immediately if there is blood or mucus in faeces, or if there is unexplained diarrhoea and wasting in growing or adult pigs, due to concerns around the disease swine dysentery.

Have you seen the pig urinate, and if so, was the urine passed in a constant stream? Intermittent squirts are not normal. If you are worried that your pig isn't urinating, then place it on a dry concrete floor and observe it for a few hours.

Changes in Skin Structure

The hairless skin of many pigs allows us to spot signs of disease such as rashes and sores that can't be seen in other farmed species. Look for signs of injury and scabs. If the pig is itching, assess whether this is excessive. Look for other pigs that may be itchy, and notice if any posts or fencing have been damaged because they have been used as a scratching post. Inspect the hair close to the body to see if you can see lice eggs, and check the ears for any crusts that could be due to mites. Check for any bite wounds on the tail or on the legs, or marks on the body that could be from serious fighting. Take photos of any blotches, rashes or strange skin so that you can show the vet, and be sure to report red blotchy skin that could be due to swine fever.

Changes in Temperature

Equip yourself with a rectal thermometer and train your pigs to be amenable to having their temperature taken. With their head in a feed bucket, get them used to a rectal thermometer being carefully introduced. Make sure that the thermometer tip is lightly touching the wall of the rectum, as taking the temperature of a lump of faeces will give an artificially low reading! A reading of over 39.5°C in a sick pig is considered high. Rectal temperature can be falsely raised if it is a very hot day, especially where there is not adequate shade and water. Be aware that a low reading (under about 38°C) will occur eventually with very serious disease as the body starts to shut down, or due to hypothermia.

The correct placement of the rectal thermometer. Note how it is held against the wall of the rectum, not straight ahead where it could sit in faeces and give a falsely low reading.

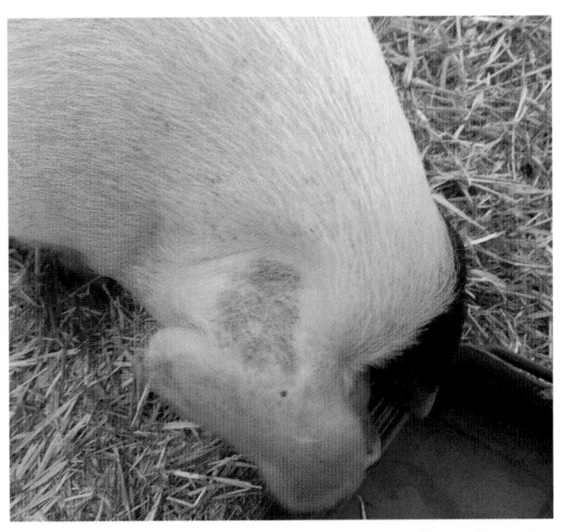

Usefully, it is often quite easy to see when there is something wrong with a pig's skin.

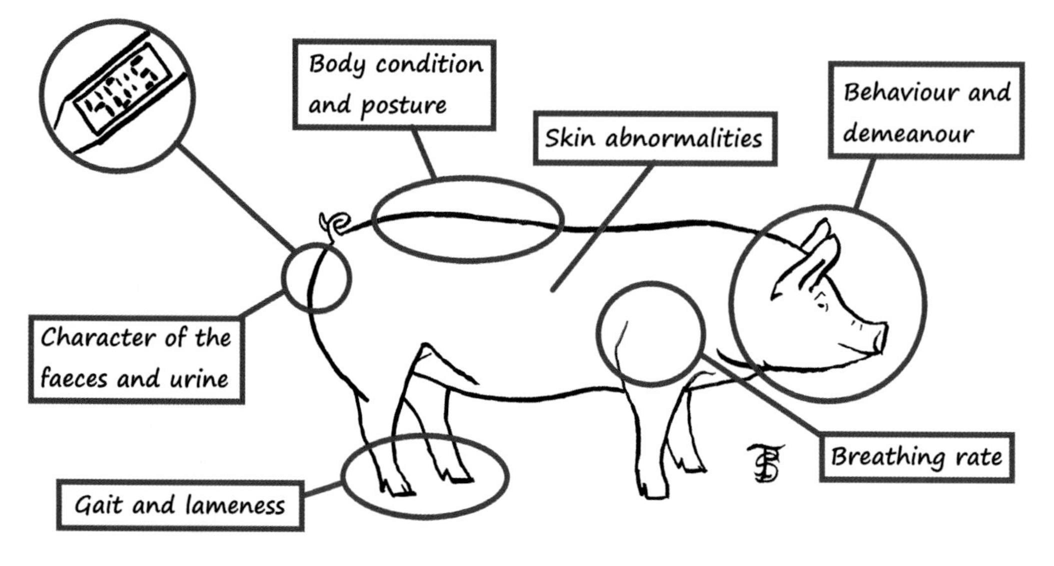

Every aspect of a sick pig should be examined and considered.

WHEN TO CALL YOUR VET

Veterinary problems range from those requiring an urgent phone call (such as a pig having a seizure), to those that may benefit from just a conversation with your vet (such as an elderly sow putting on weight).

In between these two extremes are the other maladies and conditions that affect pigs, for which the response will depend on your own experience. If you have successfully dealt with a particular condition before, and you and your vet are confident that you can recognise it, enact the treatment that has been prescribed and provide aftercare, then it may be reasonable for you to complete this without veterinary assistance, but very much under veterinary supervision. Standard operating procedures for any such instances should be detailed in your veterinary health plan.

However, when dealing with something new and unknown, you must seek professional advice before proceeding. Furthermore, always contact your vet if your normal course of action for a clinical sign has not led to a normal result. Many diseases can look similar despite having very different causes and best

Unfortunately, this very dull-looking pig could be sick for all manner of reasons.

treatments. Remember, whether or not a vet is required, it is your legal duty to attend to that sick pig straightaway, under the Animal Welfare Act.[1]

When deciding whether a vet is required immediately, it will be important to use a physical examination to inform your decision. Of course, if a pig is dying or in distress then you must phone the vet straightaway. You must also (by law) act immediately if notifiable disease may be present, and therefore it is important to recognise the signs of these diseases as described in Chapter 9, 'Diseases of Pigs'.

If you are unsure, your vet will be happy to listen to your thorough assessment of the pig and make a decision whether the call is urgent, or if it can wait until Monday (as these things are always at the weekend and are most commonly late at night). Remember that pigs are very stoic, because they are hard-wired to hide pain so they aren't picked off by predators. Therefore a sick or injured pig is likely to be in far more pain than it is demonstrating.

It would be impossible to go through every clinical sign that requires urgent veterinary attention, but the following list provides some red flags:

- A pig struggling to breathe, possibly with its mouth open
- Unexplained diarrhoea and wasting
- A male pig that is straining to urinate, can't urinate, or is dribbling urine
- A panting pig with a hunched posture
- A pig that is very off colour and is not eating, especially if the rectal temperature is high
- A pig that doesn't put weight on a limb, is holding a limb at an odd angle or a very painful lameness
- Neurological signs such as shaking, twitching, fitting, pressing their head against objects or avoiding light
- Blisters just above the snout or on the hoof
- Severe blood loss
- Acute changes to the skin such as red blotches
- Farrowing emergencies, as explained in Chapter 13, 'Breeding'
- Problems with suckling piglets, as their state of health can deteriorate very quickly
- Sudden death

When you call your vet, work through the following list so that the information is communicated in a clear and systematic way:

- Sex of the pig, neutering status, age (if known), breed (if known), use (such as a breeding sow or pet)
- Vaccination status (timings of last vaccine and vaccines used) and any diseases present on the holding, if known
- Clinical signs (using the physical examination checklist as a framework)
- Length of time showing signs
- Progression
- Any medications or interventions already taken, and the result

The vet may also want information such as recent pig movements on and off the holding, and the length of time that you have owned the pig.

ACUTE VERSUS CHRONIC

In medical terms 'acute' means sudden onset and 'chronic' means a longer duration. A broken leg is an acute injury, whereas gradual weight loss is a sign of chronic disease.

HOW TO STREAMLINE THE VET VISIT

To prepare for the vet visit, think about what your vet is likely to need to do, and what you can do to make this easier. Generally vets charge by the minute, so everything you can do to prevent wasted time standing around will help everyone. That being said, a couple of extra minutes to ensure that a thorough examination has been completed, or to make sure that everyone is fully on the same page in terms of treatment for a condition, can save unnecessary confusion and visits later on.

Get the pig into a space where it can be examined. This process can be made easier by training pigs to be amenable and quiet to handle. Pigs are incredibly intelligent animals, and like dogs, can be taught to tolerate veterinary intervention. Especially for pet pig owners, Grisha Stewart's cooperative care techniques[2] can be practised, using non-meat treats, to make everyone's experience much less stressful. A badly handled pig will give your vet few options in terms of restraint, and may leave them only with options that you are not happy with, for example snaring.

Also, be aware that a pig in pain is likely to be very stressed, so may not 'perform' exactly as per their training.

Remember that good light will be necessary, as may buckets of clean water. Do not move a pig that can't get up, or one that

This handling system, which pigs have been trained to feel comfortable to enter, will make the vet's job on this smallholding much more straightforward.

is in too much pain, and ask the vet to come to them. If the vet is coming out to complete a procedure that may require sedation, be sure to understand whether you should be starving the pig and any other preparatory steps. Try to have someone on hand who could help the vet if required, for example with restraint.

Once the vet arrives, be sure to tell them where to park. This should be balanced in terms of biosecurity (remembering that they drive on to other farms) and convenience for them to access their kit and equipment. Prepare a clean boot dip ready for them to dip their boots.

Throughout the visit, don't be afraid to ask questions. It is really important that you understand what is going on. Livestock are full of surprises even for the most experienced keepers, so the more you can communicate with farmers, vets and fellow smallholders the more proficient and self-sufficient you will become.

If at any point you are unhappy or confused, tell your vet. This includes ensuring that you understand what you are required to do in terms of treatment and aftercare.

Before your vet leaves make sure that you understand the following:

- Their thoughts on a most likely diagnosis and prognosis
- The details of any medication that you will need for your medicine records
- Anything that you are required to do after the visit
- The pig's expected progression after the visit
- What to do if the situation does not progress as they would expect
- When and if another visit should be arranged

ADMINISTERING APPROPRIATE CARE

First Aid

In the case of a true emergency, anything you can do to help before the vet arrives will be of benefit. If you are familiar with first aid in humans, then the principles of 'Airway, Breathing and Circulation' are exactly the same in animals. Try to get the pig in a comfortable position, ensure the airway is unobstructed, and if something is bleeding - put pressure on it!

Supportive Care

There are many other cases where veterinary intervention is required the same day, but it may not happen in the same hour. In these cases, attending to the pig's basic needs before the vet arrives can be hugely beneficial.

If it is possible without causing further harm, walk the pig into a safe, cool, dry area. Provide food and water within easy reach and a deep bed of straw to keep it warm. If food and water are being refused, see what else you can tempt the pig with. This is especially important for lactating sows, which will deteriorate very quickly without the sustenance needed to produce very large volumes of milk. Apple juice, mashed banana and warmed porridge oats can all be really beneficial, making sure, as always, that nothing has been through a kitchen.

Water and hydration status is particularly important for sick pigs, and often sick pigs will die of dehydration rather than due to the actual disease. For pigs that can't drink, alternative hydration provisions must be provided. This needs to be appropriate for the condition of the pig. For example, it would be dangerous to syringe water into the mouth of a pig that can't swallow, as it will likely choke. Instead, you can give small amounts of water

A pig that will not drink can be rehydrated by syringing water into the rectum.

at a time rectally by gently inserting water in a syringe several times per day. For pigs that can swallow, syringing apple juice into their mouth can provide a much-needed hydration and sugar hit.

Before giving medications to a pig that a vet is coming to see, check that this is appropriate on the phone. Even our seemingly benign pain-relieving medicines can be dangerous to give in a pig that might have a stomach ulcer, so real harm could be caused by acting outside your vet's recommendations.

Aftercare

Following the vet visit, a sick pig may require extensive aftercare. We do not have an intensive care unit for pigs, and instead, we often require pig keepers to replicate an intensive care facility as much as possible on their holding. Specifically, pigs that can't stand (for example those with meningitis) will require round-the-clock care.

Before taking on this huge responsibility, ensure that you understand what is required of you from your vet. Together, write out a care plan for the pig. We have written an example below. If this can't be fulfilled completely, it is kinder to put the pig to sleep rather than prolong suffering. This decision will be informed by many factors – for example your work routine and whether you will be able to attend to the pig every few hours, or whether you live on site to be able to give round-the-clock care. In addition to this, always make sure that you have discussed set points, after which it would be inappropriate to continue. Consider the ethics of allowing days of suffering for sick pigs. Even where the pig is likely to make it through, is the short-term suffering fair?

And finally, for producers, remember that in order to send a pig to the abattoir, all withdrawal periods for medicines must have elapsed. If this is your goal, the pig will need to improve to the extent that treatment can be stopped and withdrawal times elapsed whilst the pig is still fit for transport to the abattoir. This is rarely sensible ethically or economically.

THE HOSPITAL PEN

Any sick pig that is being kept on the farm for treatment should be housed in a 'hospital pen', which should meet the following requirements:

* House the pig so that it won't be harmed by other pigs, and where infectious disease won't spread through the holding. This may mean housing it separately (for example in the case of a rectal prolapse, which other pigs would bite off), or with a companion in other cases.
* Provide a warm and well bedded area. Small pigs may require extra warmth.
* Provide food, water and treatment as required. Water should only be withheld if a pig is likely to accidentally drown itself in it. For example, it should not be withheld from piglets with diarrhoea.
* The hospital pen should be an area that you can attend to frequently and easily. It should be well lit.
* Sick pigs should be assessed at least daily, but often much more frequently depending on the complaint.
* Pain relief (under veterinary supervision) will often be required.

Remember to take records of medicines used and the progression of disease, as these can be helpful to look back on. After a few days, several days will blend into one.

If the pig is not improving as your vet anticipated, further action must be taken, which may include euthanasia. It is really important to keep your vet up to date with the progression, and to constantly reassess the situation. Just because you have been working tirelessly to save a pig for three days does not mean that you should continue your efforts.

EXAMPLE OF A 'DOWN PIG' CARE PLAN

* Turn every six hours (to avoid nerve and muscle damage on one side)
* Check water and food every four hours. The pig should be drinking at least __ltr water per day; if not, give __ rectally
* Give medication daily
* Send an update, including a video, to the vet daily
* By two days' time, expect ___. If not, call the vet
* By four days' time, expect ___. If not, call the vet

If Things Don't Go To Plan

As spoken about extensively, where things do not go to plan, euthanasia must be considered and we encourage you to have set plans in place for this to ease stress on an awful day. As discussed in Chapter 16, 'Pig Death and Euthanasia', this must also include carcass removal by an approved fallen stock collector.

Especially for pet pig owners, referring the case to a farm veterinary hospital may be appropriate. Several UK veterinary schools have farm animal referral centres attached and are happy to deal with anything routine as well as the unusual. Be aware that not all hospitals can take cases on an emergency basis, so speak to your vet and consider your options early if this might be something you want to do.

And finally, if a pig has died, or you have had to make the really difficult decision to euthanase a pig, please do consider a post-mortem by your vet or by submitting the pig to a farm animal post-mortem service (your vet will be able to tell you more). Understanding why a pig has died may help you prevent a similar fate for your other pigs.

THE VETERINARY HEALTH PLAN

A veterinary health plan is a document that ties together animal health management on a holding and gives a framework for the keeper to follow over the next year so they might prevent and act on clinical signs of disease. A good veterinary health plan is created between the vet and the client in collaboration, and is bespoke to each holding. It is in a format that the keeper will constantly refer back to, and may even display around the holding. A visit to create a veterinary health plan gives the opportunity to discuss current practices and seek improvements where these are warranted.

The veterinary health plan may discuss the following:

- A summary of production and performance, disease and abattoir records for the last period so that these can be discussed with your vet and improvements sought if warranted. Over-inflating the figures leads to missed opportunities to improve performance, so is only cheating yourself!
- Standard operating procedures for the holding relating to biosecurity, storage of medicines, health monitoring, and so on. This is your opportunity to check with your vet that everyone attending to the pigs is on the same page, and that there is nothing you could be doing better.
- A preventative health care plan. This can be displayed in a calendar format that you can refer to throughout the year, but must be tailored to your specific holding.
- A table of conditions seen on the holding for which a treatment plan has been agreed. The clinical signs expected, prescribed treatment and expected progression should be detailed, along with any medications that you are permitted to keep on the holding, and how they should be used. Your vet can only leave medicines on site if they are content that they will only be used as prescribed. It is therefore really important that you demonstrate to them the animal handling, practical skills and measured approach that this requires. If you do not know these skills, the veterinary health plan visit is an excellent time for some training (including refresher training) with your vet.
- Trigger points that would necessitate a call to the veterinary practice. For example, particular clinical signs, or a disease not progressing as expected after treatment. This way, both you and your vet are on the same page as to when contact is expected.
- A health management action plan with future goals and hopes for the holding. These can be reviewed yearly and allow keepers to move forwards with their smallholding dreams. These could be production aims, medicine aims (for example lower antibiotic use) or even construction aims.
- A record of a contingency plan for emergencies such as fire, notifiable disease or flooding.
- A record of procedures that keepers are trained in and competent to perform, such as administration of injections or ear tagging.

The plan should be reviewed at least yearly, or when specific problems arise. There are templates available, but your vet will probably have a template that they like to use. Remember that the process will be much less

A template veterinary health plan that Pete uses for his clients at Synergy Farm Health.

useful if it is viewed as a tick-box exercise, or if it is not tailored to your specific holding.

You may be eligible for financial assistance to complete this if you keep more than fifty pigs through the Animal Health and Welfare Pathway. More information can be found at https://www.gov.uk/government/publications/animal-health-and-welfare-pathway/animal-health-and-welfare-pathway

If you would rather not use a vet to do this, or if this is something that your vet does not provide, the example below could be tweaked to your own personal situation.

Farm Name and Address	***Main Contact Name and Address***		
CPH number	xx/xxx/ xxxx		
Number of pigs			
Breed			
Organic	Y/N		
Red Tractor	Y/N		
Closed herd	Y/N		
Veterinary practice and emergency contact number			
Short term aims/ targets for the next twelve months			

Farm Name and Address	***Main Contact Name and Address***		
Longer term aims/targets			
Fertility data	Current (2022)	previous (2021)	target
Sows bred			
Gilts bred			
Average litter size			
Pigs born/sow/yr			
Piglets born			
Piglets weaned			
Pigs weaned/ sow/yr			
Assisted farrowings			
Abortions			
Stillbirths			

Disease	***status***	***Action agreed***
Erysipelas		
Parvovirus		
Worms		
Mange		
Lice		
Other		

Biosecurity plan			
Vaccine:	When to give	Which pigs	Route
Parasites	Status	Control Plan	
Internal			
External			

A CLOSER LOOK AT PAIN AND PAIN MANAGEMENT IN PIGS

Under the five freedoms,[3] discussed in Chapter 1 of this book, you have a responsibility as a pig owner to ensure that your pigs are free from pain. However, pigs are prey species, meaning that they will suppress the appearance of pain to avoid being picked off by predators. Therefore, pain can be challenging to identify in pigs. That being said, pain-free pigs are not only happier, displaying more natural traits such as searching for food and wallowing, but will also be more productive.

Pain results from tissue injury, whether that be chemical, mechanical or thermal. Receptors in the affected tissue then inform the brain of this injury via nerves. Finally, the body reacts to protect the injury. This can be demonstrated to us through clinical signs, for example a pig not wanting to bear weight on an entire limb due to pain in one joint.

Persistent stimulation of pain receptors leads to a lowering of the pain threshold over time, therefore a pig can appear in less pain even when the injury is still present. Young animals tend to appear more reactive to pain as they haven't yet learnt how to hide this.

Like us, pigs have a memory of pain that will make them unlikely to repeat painful procedures (hence the use of electric fencing), but this also means that they can take time to use a limb that is no longer painful but was at some point in the recent past.

Behavioural responses to pain include the following:

- **Changes in posture:** The pig may shift weight away from the painful area and is more likely to spend time lying down.
- **Lameness:** As above, the pig may appear lame to minimise weight bearing on the affected area.
- **Activity:** The pig may reduce its social interactions and food intake, leading to weight loss and a change in the social hierarchy for the injured pig.
- **Odd behaviours:** Acute pain will lead to a more active pig initially, which will be followed by depression. This can manifest as aggression, nosing, chewing, licking and playing.
- **Facial expression:** Facial expression and 'grimace scoring' can be used to identify pain in pigs. Wide eyes, a long snout and a stretched neck are the most reliable indicators of pain shown by pigs during farrowing.[4]

Sources of Pain

Pigs undergo uncomfortable procedures, some of which are essential for husbandry reasons, such as ear tagging, slap marking and castration in pet pigs. We can predict the occurrence of these procedures and try to help the pig in some way. For example, for slap marking we can provide the pig with a bucket of feed so they notice the procedure less, while for castrations we can give pain-relieving medication.

Pain is also associated with normal life processes such as giving birth, which is generally accepted as a painful process in all species! The harmful effects of this pain must be considered, and keepers must work hard to ensure that the sow is eating and drinking as quickly as possible after birth. You may also want to provide these sows with pain-relieving medicines, so do talk to your vet about this possibility.

Disease is the final cause of pain. We must be constantly looking for indicators of pain and strive to understand the cause. If pigs are exhibiting the signs mentioned above that are consistent with pain, please phone your vet so that you may discuss the most appropriate course of action. We must also strive to prevent disease and the resultant pain, which can be done by following the advice given throughout this book.

Lastly, it is essential to remember that all pain-relieving medicines in pigs can only be prescribed by a veterinary surgeon, and should only be given under the direct guidance of your vet.

CHAPTER 9

DISEASES OF PIGS

This chapter is not to take the place of a veterinary surgeon. The diseases mentioned are to make you aware that they exist, so that you are more likely to spot them and understand how they may present in your pigs. It is important to note that it is illegal for anyone who is not a veterinary surgeon to diagnose a disease in animals or to give advice based upon a diagnosis.[1]

If you suspect one of these diseases, please contact your vet to inform them of the clinical signs that you have noted, including their duration, the time you have had these pigs on your holding, where they originate from, and what vaccinations they have had. Do not try to diagnose these diseases without veterinary supervision, and certainly don't attempt to treat them. By contacting your veterinary surgeon as soon as possible not only will you limit spread within your pigs, but hopefully the spread of disease to neighbouring holdings.

Firstly, we will discuss the broad categories that diseases can fit into. Then we will tell you about two types of disease that are particularly important to watch for: the notifiable and zoonotic diseases. Finally we will go through each body system and tell you about some of the most common diseases relevant to smallholder pigs.

To learn more, the book *Pig Diseases* by D.J. Taylor[2] is a fairly low cost, comprehensive disease guide, which has greatly informed this chapter.

BROAD DISEASE CATEGORIES

Non-Infectious Diseases

Many diseases of pigs have non-infectious causes. Non-infectious disease signs are often caused by something not suiting the pig, such as its diet or environment, which leads to a structural issue. This is good news, as with some management tweaks we can often improve the prospects of the other pigs on the holding, or hopefully stop the progression of disease.

Cancer is a non-infectious cause of disease where prognosis is unlikely to be impacted by management changes; luckily it is quite rare in all but our eldest pigs.

Infectious Diseases

Other diseases are caused by pathogens. These are microorganisms that cause disease, such as bacteria, viruses and parasites. Some of these will be particularly dangerous due to their potential to spread across holdings, and all pig keepers need to be aware of the clinical signs associated with them. Some of these diseases are not currently present in the UK and their introduction would be disastrous for us.

It is important to note that not all microorganisms are a risk to our pigs. Nearly every surface of a pig (its skin, its intestines, its respiratory tract) is covered in a rich flora of bacteria, viruses and even parasites. In general, these work in harmony together and keep the pig safe. For many pathogens, it is only when the balance of microorganisms is upset that pigs experience disease.

Disease-causing pathogens can be broadly split into the categories described below.

Bacterial Diseases

Bacteria are simple, single-celled organisms. They contain DNA inside a capsule and can survive independently. Bacteria vary in their shape, their structure, where they can live, the mechanisms that they use to survive, and more. This means that they vary hugely in the way they cause disease - or not, as the vast majority of bacteria are completely harmless.

In terms of treating bacterial disease we think of antibiotics as our greatest weapon. However, it is important to notice the huge variety of bacterial pathogens. Due to this variation, not every antibiotic can treat every bacteria! Therefore, understanding exactly which bacteria might be at play is crucial in order to pick the correct antibiotic, should one be necessary.

Viral Diseases

Viruses are even smaller than bacteria. They don't possess the necessary structures needed to replicate, so are dependent on host cells for this. However, they can still survive outside cells and hosts, and some can remain viable for long periods of time on surfaces. They can also spread in rather extreme ways - for example, viral plumes caused by foot-and-mouth disease outbreaks, or porcine respiratory and reproductive syndrome (PRRS), can cause a virus to spread over several kilometres.

Not all viruses kill animals, but they do make it much more likely for bacteria to infect and cause disease to organs due to the destruction of an animal's immune defence systems. All viruses will increase the chances of other pathogens, such as bacteria, infecting a pig.

Viruses are not treated by antibiotics, and generally we have no treatment for them other than supporting the pig through their disease as much as possible. Therefore prevention is key!

Parasitic Diseases

Parasites are microorganisms that live in, or on, a host and feed off it. They are much larger than both bacteria and viruses. Some are single-celled, such as coccidia, others are multicellular, such as parasitic worms. In pigs, parasites cause a range of clinical signs, from slightly reduced growth rates to deadly disease.

Parasites are treated with antiparasitic medicines, and different categories of parasites are killed with different antiparasitic drugs. Parasites are broadly grouped into those that live inside the pig (internal parasites), and those that live on the pig (external parasites).

Fungal Diseases

Examples of fungi are mushrooms, moulds and yeasts. They are less common causes of UK pig disease and therefore the only example that we will cover is the skin disease ringworm. It is also important to note that the toxins that cause mycotoxicities in pigs have been produced by moulds.

Zoonotic Diseases

Out of these infectious diseases, there are some that can be passed from pigs to humans (and humans to pigs). These are called zoonotic diseases. For most of these, the people most at risk are the young and immunocompromised.

We have covered sensible precautions around zoonotic disease in Chapter 3 'Biosecurity'. If you display any signs of disease you must notify your doctor of the animal species with which you are in daily contact, and any clinical signs they are exhibiting, especially if they are consistent with your own.

Diseases that can be zoonotic are highlighted in the disease summaries throughout the chapter. Thankfully the UK is currently free of two possible pig zoonoses, brucellosis and trichinellosis, so these are not discussed.

In terms of food safety, we must consider the risk from eating insufficiently cooked pork meat, particularly regarding Taenia solium, which is a tapeworm known as pork worm. When humans are infected a cyst can form in the brain, and most cases present as epilepsy.

NOTIFIABLE DISEASES

What is a notifiable disease? This is a disease that, by statutory requirements,[3] must be reported to public health or veterinary authorities when it is suspected, due to its importance to human or animal health. Notifiable diseases are often capable of very fast spread. This means that prompt diagnosis is imperative so action can be taken and hopefully its spread halted.

Once identified, whole herd culls and trade restrictions are often implemented to try to control the disease, which would be catastrophic for both the UK pig herd and for small-scale keepers.

If you suspect notifiable disease on your holding, please phone your veterinary surgeon, who will then inspect the animals to decide whether they need to report the disease to the Animal and Plant Health Agency (APHA). Alternatively, you can phone the Defra Rural Services Helpline on 03000 200 301, or if you live in Wales then contact 0300 303 8268. In Scotland you should contact your local Field Services Office.

Failure to report a notifiable disease is a criminal offence. The Animal Health Act 1981[4] says that pig keepers must be familiar with the clinical signs of notifiable disease to be able to report it when the signs are first seen. Therefore we hope that the following descriptions will allow you to meet these requirements. The list of notifiable diseases is fluid and may change within a day, so please don't consider this list exhaustive.

African Swine Fever (ASF)

ASF has not yet been seen in the UK, but it will arrive on these shores one day. The disease continues to move west from Europe, where it has been found as close as Germany. It has also decimated the Chinese pig industry, wiping out about half of their pigs.[5] In domestic pigs it is highly infectious and highly fatal; however, wild boars often show much milder disease. For this reason, the wild boar UK population is a specific worry in terms of introduction of this disease into the domestic British pig herd.

The disease is most likely to enter the UK through infected meat or pork products, and is most likely to infect a pig through inadvertent feeding of infected meat.[6] The authorities have already detected meat infected with ASF in the UK, by testing meat illegally imported into Northern Ireland.[7] Therefore, we can be confident that it is in someone's freezer: we just need to make sure that it doesn't get into our pigs. The ASF virus can survive both freezing and curing. Once entered, infection will spread through pig-to-pig contact, or via contaminated people or objects moving between holdings.

ASF most commonly presents acutely.[8] Pigs may suddenly die, or may have a high fever (>41°C) with red or purple skin discolorations. Pigs will huddle and be reluctant to move or eat. Other signs may be present, for example vomiting and diarrhoea (possibly bloody), loss of balance, coughing and abortion.

When first introduced, ASF may cause these signs in just one or two pigs while the rest of the herd become infected.[8] Therefore it must be considered in every very sick or suddenly deceased pig.

Both ASF and classical swine fever (*see* next entry) can look indistinguishable on post-mortem examination from a common disease caused by porcine circovirus type 2 (PCV2). Therefore, suspicious cases of PCV2 must also be reported as potentially notifiable.

Please visit The Pirbright Institute's website (https://www.pirbright.ac.uk/asfv) to look at some photos of the condition to ensure you are familiar with the clinical signs.

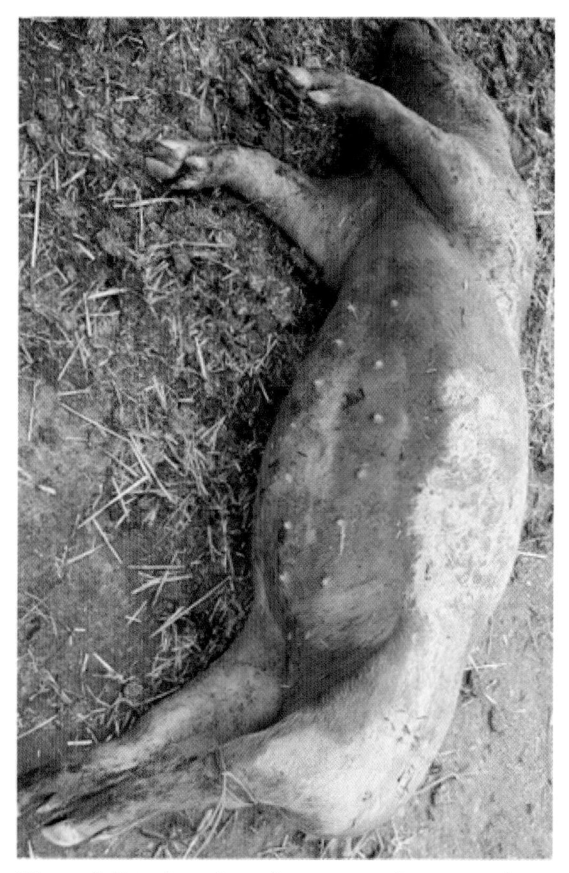

The red discoloration of a carcass that caused Claire to be worried about ASF or CSF. Luckily, with further investigation notifiable disease was ruled out.

Classical Swine Fever (CSF)

The two swine fevers are caused by very different viruses, but are paired due to their similarities in clinical signs. CSF (also known as 'hog cholera') was last seen in the UK in 2000.[9] It circulates continuously throughout much of Europe, Asia and South America. Similarly to ASF, infected pork products are our greatest risk for entry into the UK. It can survive for four years in frozen meat and three months in salted meat,[10] which indicates why it is so important that kitchen waste is not fed to your pigs! The virus spreads from pig to pig through body fluids and secretions, which could occur directly or through contamination of objects and people.

There are several different strains of CSF, meaning that the seriousness of clinical signs can range from fatal disease to very few clinical signs.[9] Where observed, clinical signs are similar to ASF with high fever (more than 41°C), loss of appetite, lethargy and sudden death.[9] Vomiting, diarrhoea or constipation, red or purple skin discolorations, coughing and difficulty breathing can all be seen, as well as abortions, weakness, tremors and convulsions in piglets.

Foot and Mouth Disease (FMD)

The FMD outbreak in 2001 is still something many of you will remember; the pyres once seen either by the eye or in the media are not forgotten. The first case of FMD was spotted in a pig at slaughter in Essex, but had been transported from Northumberland. The owner of the infected pigs was found guilty of not reporting a suspected notifiable disease to the authorities, and was banned from keeping pigs for fifteen years.[11] The owner was also found guilty of having fed 'unprocessed waste' to his pigs, leading to their infection. This led to the ban of feeding 'pig swill', which is why we can no longer feed untreated kitchen waste to pigs. We do not vaccinate against the disease in the UK, meaning that as a trading country we maintain our FMD-free status.

FMD is a highly contagious virus, with often 100 per cent of pigs affected. Very few adults die from the disease, but mortality is up to 20 per cent in piglets. Pigs are amplifiers of the disease, meaning that they produce a large amount of virus, which will then infect pigs, cattle, sheep and goats. The virus can be transmitted either directly from farm animal to farm animal, indirectly through objects and people, or in exhaled air.

Pigs develop a sudden and painful lameness, squealing loudly.[12] They will want to lie down, and will not want to feed. Generally, pigs don't develop the blisters that we associate with this disease in other species. In some cases blisters may be observed on the skin just above the hoof and on the snout. Blisters in the mouth are rarer.

Swine Vesicular Disease (SVD)

We cannot talk about FMD without mentioning SVD as the two diseases are indistinguishable from each other in how they present in pigs, so any clinical signs of SVD must be reported so that testing can be done to rule out FMD. The main clinical sign is blisters (vesicles), which appear at the top of the hooves, between the toes, and more rarely on the snout, tongue and lips.[13] These blisters lead to lameness, lethargy and loss of appetite.

The disease is spread via pigs contacting infected fluids, or by pigs eating contaminated meat sources. Again, it is vital for you to disinfect your vehicles, housing and yourselves to prevent transmission between holdings.

Aujeszky's Disease (Pseudorabies)

Aujeszky's disease was last seen in France in 2019[14] and can infect our traditional farmed animals as well as dogs, cats and rats.[15] The virus responsible is a herpes virus, which, just like a cold sore, is easily transmitted via close contact.

The clinical signs observed in piglets include shivering, incoordination and weakness of the back legs.[15] Adult pigs may experience difficulty breathing, a fever, weight loss and abortion. Conditions that may look similar to Aujeszky's include salt poisoning/water deprivation and *Streptococcus suis* meningitis, which will be discussed later in the chapter.

Teschen Disease

Teschen disease has never been seen in the UK but also presents with neurological signs.[16] Affected pigs will have a fever, be lethargic, and will be unable to coordinate their movements. They may also tooth grind, convulse, lip smack, be unable to vocalise or be paralysed.

Porcine Epidemic Diarrhoea Virus (PEDv)

This disease is caused by a coronavirus that infects the gut lining causing a watery diarrhoea and dehydration.[17] Pigs may also have a fever, and may vomit and stop eating. New-born piglets usually die, but older pigs experience less severe disease and often lose weight. Infection is spread through infected faeces and therefore strict biosecurity is imperative for this disease.

If there is diarrhoea affecting a group of animals, it is crucial to discuss this with your vet. There are further diseases such as swine dysentery or transmissible gastroenteritis in pigs that would also be deeply worrying, despite not being technically notifiable. Therefore it is really important that we get to the bottom of this clinical sign.

There are many excellent AHDB resources on PEDv and keepers are encouraged to sign up to the significant diseases charter,[18] which covers both PEDv and swine dysentery.

Tuberculosis (TB)

Pigs very rarely contract the zoonotic disease bovine TB, but as the disease continues to spread throughout the UK in cattle and our wildlife population, the disease can spill over into less typically affected species.[19] Due to the importance of wildlife spread in this disease, outdoor pigs are at particular risk, especially in the TB hotspots in the south-west of England and Wales. Measures that prevent badgers and other wildlife species (for example wild boar) coming across pigs or their feed should be practised.[19]

A pig with TB will typically exhibit no clinical signs, and the disease is generally found post slaughter. If the disease were to progress, weight loss and respiratory signs, especially coughing, would be observed.

Anthrax

Anthrax is a bacterial infection that can also infect humans, but is luckily very rare in pigs. It leads to the sudden death of an animal, often with the fairly tell-tale sign of bleeding from the mouth and anus. You must discuss the sudden death of a pig with a vet, who will then decide whether the carcass needs to be tested for anthrax. Due to the infectious nature of this zoonotic disease, you must not cut open or move a dead pig without being sure that they have not been killed by anthrax, so please do not do so until you have spoken to a vet.

COMMON DISEASES OF PIGS

Next we will go through the commonest diseases seen in UK smallholder pigs through the bodily systems that they affect and the clinical signs that they are likely to cause. This is quite challenging to do as many diseases can affect several bodily systems. Therefore we have placed diseases with the bodily system that they are most likely to affect. We won't repeat the information already discussed in the notifiable diseases section of this chapter, so make sure, first and foremost, that you understand how to spot those.

GENERALISED DISEASE

Erysipelas

Erysipelas is a disease caused by a bacteria (*Erysipelothrix rhusiopathiae*) that is very, very common, especially in non-commercial pigs.[20] For this reason it is really important to vaccinate against this disease. Rarely, this bacteria can also infect humans in a condition called erysipeloid. The bacteria can survive in soil for up to a month.

Signs of this disease in pigs include sepsis, infection of the heart valves, arthritis, reproductive problems and skin necrosis (death). Traditionally, the most tell-tale sign in a pig were diamond-shaped skin lesions. This presentation has changed in recent

The large growth on this pig's heart valves was due to erysipelas infection.

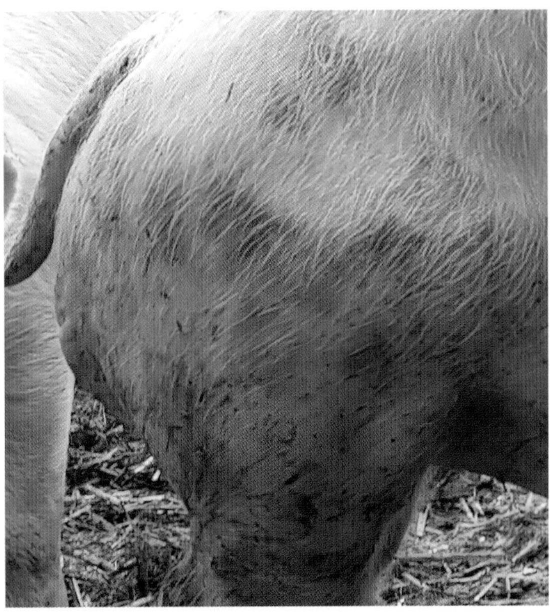

When present, erysipelas skin lesions are red/purple and raised.

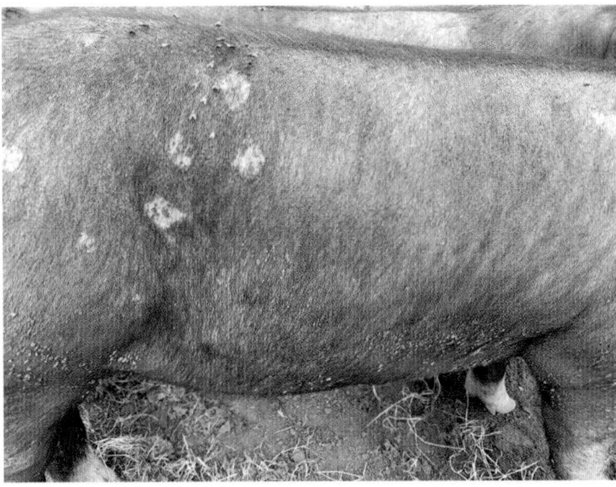

The classic diamond shape of healing erysipelas skin lesions.

years, meaning that we often do not see any skin lesions and instead most commonly see sudden death or general sickness. Many cases will only be diagnosed at post-mortem by visualisation of growths on the heart valve. Pigs sick with erysipelas will have a fever (above about 39.5°C), will stop eating and may show neurological signs. The pig's ears may also start to turn blue as a sign of sepsis.

When present, skin lesions may also not be the perfect diamonds pictured, but may just be blotchy or look like a rash. Taking some photos of these to send your vet can be really useful.

If caught early, erysipelas treatment can be attempted under supervision from your vet, however this may require long courses of medication. If disease has progressed to serious growths on the heart or is acute in presentation, treatment is rarely successful. After infection, pigs can also suffer with debilitating arthritis, which cannot be cured. Prevention is key to avoiding cases by vaccinating all pigs for this disease.

Haemophilus Parasuis (Glässer's Disease)

Glässer's is a bacterial infection that causes a range of clinical signs in young pigs. Clinical signs experienced are due to sepsis, as well as

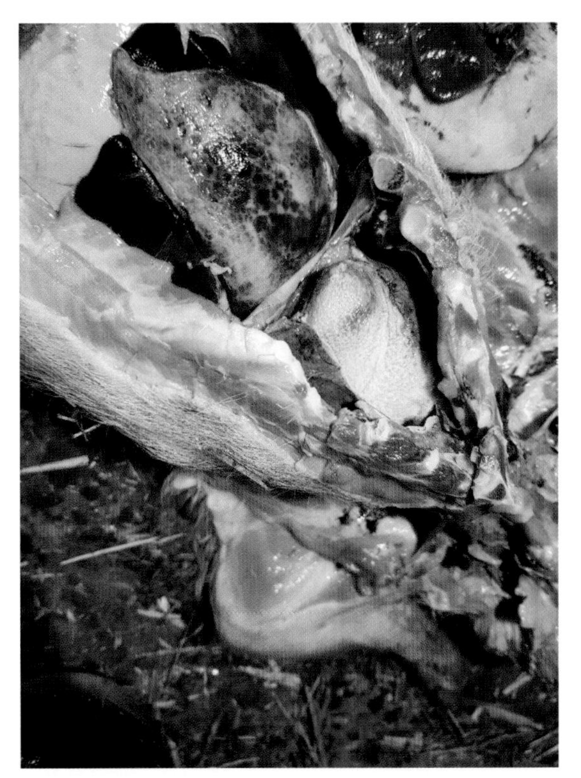

The sudden death of this pig was due to thick 'glassy' deposits over the liver and heart after *Haemophilus parasuis* infection.

thick 'glassy' deposits that form on internal surfaces, such as the meninges (brain coatings), lungs, joints and abdomen. Even if a pig survives this infection, these deposits can cause serious problems later in life.

Affected pigs are generally three to six weeks old (but may be older), and develop difficulty breathing, and fever. Pigs may cough and become lame with joint swellings. As the disease progresses, pigs will be unable to stand. They will lie on their belly and extend their neck, struggling to breathe. Meningitis tends to occur later in the disease process, or in older pigs with sudden death. Respiratory signs can allow the disease to be distinguished from *Streptococcus suis* during the pig's life, and disease can be confirmed with post-mortem and laboratory sampling. Vaccination can be used to prevent this disease.

Iron Deficiency Anaemia

Suckling pigs have a high need for iron; they receive around half of their requirement through milk, and are therefore at risk of developing iron deficiency anaemia. They can obtain the rest from soil, however indoor farrowing practices will disrupt this.

A large tub is ideal for temporarily containing injected piglets to prevent any getting a second dose of iron.

Affected piglets of one to three weeks of age become pale, lose weight, develop rough hair coats and have diarrhoea. Severe cases will show laboured breathing and listlessness, and will die. Be aware that blood loss (possibly internally) will cause a very similar set of clinical signs.

To prevent the condition, indoor producers will inject three-day-old piglets with iron. Some smallholders will bring some good quality soil into the pen for piglets to root around in, which is excellent for both iron and enrichment.

Piglets with vitamin E deficiency can be susceptible to an iron toxicity, meaning that piglets deficient in vitamin E may die if they are injected with iron. Therefore, look out for the signs of iron deficiency anaemia and speak to your vet about whether iron supplementation may be appropriate for your holding.

Vitamin E/ Selenium Deficiency (Mulberry Heart Disease/ White Muscle Disease)

This is a disease of rapidly growing pigs (roughly two to six weeks of age) that are fed a diet that is low in vitamin E and selenium, such as cereals. All muscle tissues are affected by the deficiency, but the high-functioning demands of the cardiac muscle mean that it is the first to stop functioning, generally causing the piglet to die suddenly after a stressful event. Several piglets may be affected at one time. The condition is diagnosed at post mortem, and should be prevented by feeding a complete diet designed for a quickly growing piglet.

Mycotoxicity

Mycotoxicity occurs when fungi colonise poorly kept feedstuffs or bedding and release

This poor pig had likely suffered from some kind of mycotoxity, causing her ears to die off. Due to the difficulties around testing for mycotoxin, this was never definitively proven.

This pig was getting too hot at a show and Alice successfully managed to cool him down. Surprisingly, pigs tend to really quite like this treatment, and it doesn't tend to stress them.

toxins that can make animals ill. There are many potential mycotoxins that pigs can react to, each with its own distinct set of clinical signs, including vomiting, infertility and death of tissues such as the ears.

Treatment is supportive, often consisting of pain relief and alternative feed or bedding, and the addition of dietary myotoxin binders can be appropriate in some cases. Prevention is achieved by sourcing good quality food and bedding, and storing it to remain dry.

Heat Stroke

Pigs have only a limited number of functioning sweat glands, and because they also have a thick layer of subcutaneous fat, they are really sensitive to heat stroke.[24] A pig suffering from heat stroke will have a high respiratory rate - higher than fifty breaths per minute is very dangerous[21] - and a high rectal temperature (>39.5°C). Bigger pigs are particularly susceptible.

Treatment for a pig experiencing heat stroke is to cool it down, spraying it with water if needs be. Pigs placed in very cold water may enter shock, just as a human might, so try to use tepid water. Afterwards, place the pig in a cool environment with adequate shelter and shade.

Heat stroke should be prevented using the steps outlined in Chapter 4, 'Pig Housing and Environment'.

Rodenticide (Rat) Toxicity

This common toxicity amongst smallholding pigs causes a pale and weak pig, sometimes with redness over its body from haemorrhages, and bleeding from the head and anus.[20] Pigs that have eaten rodenticide directly can be affected, as well as pigs that have eaten poisoned rats.

Bracken Toxicity

Pigs with bracken poisoning experience heart failure, which causes fluid to back up in the lungs and chest. Many cases are found dead or may be found struggling to breathe. The condition is commonly identified where pigs have access to heathland or woodland. This is also a food safety concern, and pigs must be withdrawn from bracken exposure at least fifteen days before slaughter.[22]

Bracken is very common around the UK countryside.

NEUROLOGICAL DISEASE

Meningitis

Meningitis is a condition where the casings around the brain (meninges) become inflamed. Farmers will often use the name of the bacteria *Streptococcus suis* and the term meningitis interchangeably, but it is important to note that other bacteria can be implicated, which require diagnostic testing to distinguish between them.

Meningitis is intensely and increasingly painful, as pressure builds in the pig's skull. Therefore veterinary advice must be sought immediately, and pigs should be monitored closely to ensure that cases are identified promptly. On top of the medication that will be prescribed, first aid is imperative. Down pigs should be turned, hydration must be maintained (rectally if pigs cannot swallow), and pigs should be bedded very comfortably in a hospital pen. Euthanasia is strongly advised within a couple of days of a pig not responding to treatment, or in very serious cases.

Causes of meningitis are often sparked by stress, such as a sudden change in temperature, transport, or mixing with other pigs. Therefore, minimising stress is important for prevention of these diseases.

Remember that signs of meningitis can look very similar to salt deprivation, so always check water availability when you see neurological signs in pigs.

Streptococcus suis

Several types of *Streptococcus suis* bacteria can cause meningitis in pigs, but *Streptococcus suis* type 2 is most common in the UK. The bacteria is carried in the tonsils and throat of the pig as well as the vagina of sows, and therefore piglets are colonised with this bacteria at birth. The bacteria can also move from pig to pig later in life. Disease can be observed in pigs up to finishing weight, however smallholdings tend to experience sporadic cases in piglets shortly before weaning or in the weeks shortly after. Smaller piglets that may not have drunk enough colostrum and milk are often affected.

Clinical signs range from sudden death to a pig appearing off balance and not eating. Early signs that may be detected include a piglet isolating itself from the group, not behaving in ways you would expect, and eyes flicking to one side. Middle ear infections can also result, causing the head to tilt to the side. If the disease progresses, pigs will lay on their side, paddling their legs or tremoring. *Streptococcus suis* is also zoonotic and can cause meningitis, septicaemia and arthritis in people.

Commercially produced vaccines don't tend to be used much in the UK in commercial pig herds, as none offers full protection. Some herds will have a vaccine made up of the strains found on a particular

Piglet suffering from streptococcal meningitis: note the swollen joints and wide eyes. Her eyes were also flicking and she was tremoring. However, due to excellent nursing, her keeper was able to bring her back to good health.

holding, however this is unlikely to be feasible for smallholdings. Some commercial herds have undergone an eradication programme for *Streptococcus suis* type 2, meaning that they don't see disease from this type; however, this is also likely to be not feasible for smallholdings.

E. coli Bowel Oedema

E. coli bacteria cause two main diseases in pigs; the other (as a cause of diarrhoea) will be covered later. *E. coli* bowel oedema causes neurological signs and sudden death, sometimes accompanied with diarrhoea in recently weaned pigs. The disease seems to fluctuate in how common it is; in recent years in the UK it is fairly uncommon, however it can be a more serious problem on individual holdings. The disease may become more common again in commercial herds as zinc oxide is removed from pig diets.

Clinical signs are caused by the toxins that the bacteria release. Affected pigs are either found dead, or experiencing varying degrees of neurological disease. Pigs will start by behaving strangely, appearing dull and losing balance, and eventually will deteriorate to lying on their side paddling their legs. Affected pigs will generally have a fever. Piglets will often utter a very high pitched squeal, and may have puffy eyelids and faces. The disease is not always accompanied by the diarrhoea that you might expect from an *E. coli*, but it can be.

E. coli vaccines given to the sow during pregnancy may not prevent clinical signs of bowel oedema, as immunity conferred through colostrum has often decreased by this point. Piglet vaccines are available but are often delivered orally through water, which is not feasible for most smallholdings.

E. coli can pass to humans, and therefore gloves and good hygiene should always be used when tending to pigs, especially when they are sick.

Haemophilus Parasuis (Glässer's Disease)

The respiratory signs accompanying the neurological signs experienced mean that it can often be distinguished from Streptococcus suis during the pig's life.

Also note that erysipelas can cause meningitis signs less commonly. *Klebsiella pneumoniae* subsp. *pneumoniae* (*Klebsiella* septicaemia) can also look very similar to these diseases. Cases can be seen during the summer months in the UK, affecting pre-weaned piglets.[23]

Water Deprivation/ Salt Poisoning

Water deprivation in pigs is particularly serious. Pigs will suffer if they are chronically restricted in terms of water intake, or can show clinical signs very acutely if water supply is suddenly halted. This can happen especially when pipes are frozen, burst or during very hot weather. Always ensure that electric fencing does not touch drinkers, and check pigs' water supply at least daily, but more frequently in hot weather or for lactating sows.

Initially pigs present as thirsty and constipated, and they can also be itchy. Neurological signs then develop including blindness, unresponsiveness to stimuli, head-pressing (against a wall), circling and seizures. Cases are distinguished from infectious meningitis by confirming that the water supply has been interrupted.

Veterinary discussion is required for any cases, as euthanasia may be required. Sudden rehydration can exacerbate the condition, so you may find that some pigs deteriorate further when the water supply is reintroduced.

Hemlock resembles a tasty root.

Hemlock Toxicity

Hemlock is a plant in the carrot family found throughout the UK. Poisoning leads to nervous signs such as trembling, drooling and blindness, or congenital deformities in piglets if it is ingested during pregnancy.[20] Other related plants (especially the leaves) can also be toxic, such as parsnips, carrots, parsley and cumin, but only parsnips seem to cause neurological signs. Paddocks should be regularly monitored for the presence of hemlock, and access prevented.

Modern paints do not use lead, but an old painted door such as this could be a source of lead poisoning to pigs.

Lead Toxicity

Lead toxicity is uncommon but most frequently occurs where soil is high in lead. Pigs will show neurological signs such as tooth grinding, blindness, twitching and poor balance.[20]

Nightshade Toxicity

The name 'nightshade' encompasses several species, which include black, woody or bittersweet and the famous deadly nightshade. These are all relatively common

Woody nightshade is a less poisonous relative of the notorious deadly nightshade.

and produce a variety of effects if ingested by pigs, including nervous signs, cardiovascular effects and vomiting. Black nightshade is a common weed of vegetable and spring legume crops.[24]

GASTROINTESTINAL DISEASES

Causes of gastrointestinal disease in pigs vary hugely in terms of the age of the pig. There are some causes that affect all pigs, some that only affect piglets, and some that only affect adolescent or adult pigs. Despite this, there are generally multiple possible causes for each age and sign, all have different best treatments, and they are often impossible to diagnose without laboratory sampling.

The zoonotic implications of salmonellosis and many other gastrointestinal diseases, and the seriousness of diseases such as swine dysentery, really heighten the need to know what pathogens are involved, so please contact your vet to allow a cause to be identified. Supportive care for gastrointestinal disease is imperative, especially as viruses are often implicated. Very young piglets suffering with diarrhoea are very likely to die due to dehydration or an imbalance of salts in the blood. Usefully, many of these diseases can be prevented with vaccines and improved management, especially around good biosecurity, hygiene and colostrum intake.

E. Coli

E. coli bacteria most commonly cause diarrhoea in piglets at two different time points.[2] Piglets under one week of age present with watery diarrhoea (pale yellow), dullness, and a very sorry-looking appearance that leads to death, or they may suddenly die. Piglets from about three to five weeks old present with greyish diarrhoea and look very hairy, small and emaciated. Affected pigs may die or may recover, but struggle to grow to their full potential.

As with other causes of piglet diarrhoea, urgent treatment under the supervision of a vet is required to maintain the hydration and energy status of the piglet, for example by providing electrolytes. Pre-weaning *E. coli* can often be prevented with improved colostrum intake, vaccines and improved hygiene. On commercial holdings, post-weaning vaccines are often delivered through the water. It is likely that these are not feasible for smallholdings, who instead should focus on minimising stress at this time. The later weaning on smallholdings should greatly assist this.

As mentioned previously, *E. coli* can also cause disease in humans, so always wear gloves, and operate good personal hygiene when tending to these cases.

Rotavirus

Rotavirus also causes diarrhoea in pre-weaned piglets, often in conjunction with other causes such as *E. coli*. Initially piglets show poor appetite, dullness and sometimes vomiting. Then profuse, yellow or grey diarrhoea will occur and the piglet will look very gaunt. If piglets are well supported with electrolytes to prevent dehydration or starvation, most survive. However, the presence of other pathogens may disrupt this.

Piglets are often infected from their sow's faeces, and therefore it is really important that the farrowing space is kept very clean and faeces are removed twice daily. Piglets should also be kept warm enough, and should be managed to ensure that colostrum and milk intake is adequate. Rotaviruses are very resistant in the environment so arcs should be moved between litters if farrowing outside, and all bedding disposed of.

As there is no vaccine for pig rotavirus, historically pig farmers would feed the placenta from older sows to pre-farrowing sows to encourage immunity for this disease in a process called 'feedback'. However, this is completely inappropriate. Firstly, this practice is illegal as we are not allowed to feed pig waste to other pigs. Secondly, the practice risks transferring diseases to the pig, rather than just exposing the immune system, which can be highly dangerous. Piglet diarrhoea has also been used in a similar way, however this will also mean that any diseases present in the faeces will be passed to the sow, and is therefore also inappropriate.

Clostridium perfringens

This bacteria leads to sudden death in piglets generally less than one week of age; those infected that are still alive produce bloody diarrhoea. Overfed piglets are predisposed to this disease, so be careful if tubing or bottle feeding. Vaccinating the sow pre-farrowing and ensuring adequate colostrum intake will help to prevent the disease. Clostridium bacteria are commonly found in soil, so outdoor production is a specific risk for this disease. Saying that, many indoor pig holdings are also contaminated with the bacteria, especially where hygiene is suboptimal.

Coccidia

Coccidia are single-celled parasites, and piglets are generally infected by ingesting egg stages (oocysts) from a previous litter. These eggs love high temperatures, so thrive under a heat lamp.

Yellow or grey, smelly diarrhoea occurs at seven to twenty-one days of age. Piglets can look scrawny and become dehydrated, but generally don't die. However, the piglet may die from another implicated pathogen.

Sometimes disease can be identified using the faecal egg-counting technique that we discuss in Chapter 12, 'Parasite Control'. However, eggs are not excreted during every stage so may not be found. Diagnosis is really important to avoid inappropriate antibiotic use in these parasitic cases, and instead a coccidiostat can be used to prevent or treat the disease. For most smallholders it is most appropriate only to do this following confirmation of coccidial disease, but your vet will be able to advise you on the use of this prescription-only medicine.

Because the eggs are so robust to their environment, control revolves around good

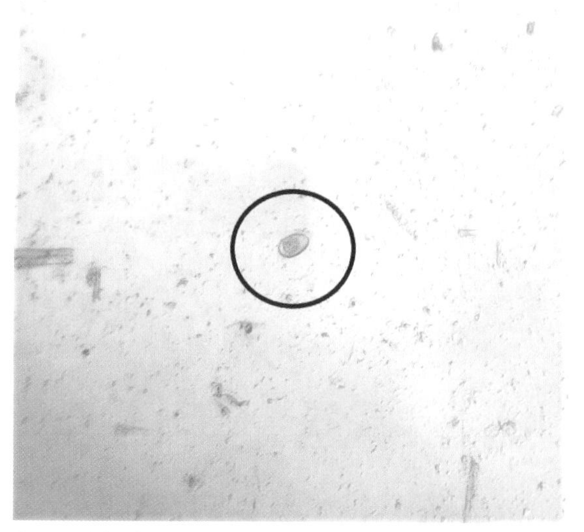

A coccidial oocyst (egg) under the microscope, provided by Synergy's talented lab technicians Abbie and Sam.

hygiene and breaking the cycle with thorough cleaning and disinfection between piglets. Eggs can survive in the environment for many months, so a fresh paddock and a thorough clean of arcs should be performed between litters. Special oocidal (egg killing) disinfectants are required for this, details of which can be found in Chapter 3, 'Biosecurity'. Rodents can be significant in spreading the disease, so control is important.

Salmonellosis

Salmonellosis describes any disease caused by the bacteria salmonella, of which there are several different types. Due to its importance as a food-borne disease, salmonellosis is reportable. This means that a diagnosis of salmonella must be reported to the APHA (by law) for further investigations. Salmonella species are carried by all other farmed species, wildlife and vermin, including birds. Carrier animals are common, meaning that pigs can carry the bacteria, infecting other pigs but not showing any clinical signs themselves.

Salmonella tends to cause three sets of clinical signs, which may depend on the age of the pig or the type of salmonella implicated.[2] Pre-weaned piglets experience watery, yellowish diarrhoea, dullness and sometimes neurological signs. Piglets one to four months old most frequently present with sudden death, or diarrhoea (often yellow) followed by sudden death. Sick piglets will have a high fever, will possibly show neurological signs, and may have blue-tinged ears (as with other causes of septicaemia). Finally, the chronic form of the disease is most common in pigs two to four months old, but can occur in any age. Pigs have weight loss due to diarrhoea and sometimes rectal strictures (meaning they can't defecate properly). Others become emaciated to the point of death.

Unfortunately no vaccination is available. Infected pigs should be isolated where

Yellow faeces in recently weaned pigs, which needs investigating to determine the cause.

possible, the presence of vermin should be minimised, and good hygiene should be practised. In terms of the risk to human health, keepers should wear gloves when handling pigs with diarrhoea, and should wash their hands before eating. If you show signs that could be associated with salmonella, please speak to your doctor.

Swine Dysentery

Swine dysentery is a huge worry to the UK pig sector currently, and smallholders have been identified as specific risks for this disease, especially the showing community. The disease is caused by the bacteria *Brachyspira hyodysenteriae*, which is spread from pig to pig and, to a lesser extent, by rodents. The disease spreads quickly and easily between herds, as the bacteria from faeces can be easily transported on pigs or on people or objects. Therefore this disease can be prevented with thorough biosecurity.

The disease is serious, meaning that some pigs will die, and it is very difficult to remove from a holding. Outdoors it is nearly impossible. Furthermore, antibiotic resistance is increasing in this bacteria, meaning that it is increasingly difficult to treat.

The disease causes diarrhoea and weight loss (or lack of weight gain). Historically diarrhoea has contained blood and mucus, however in many recent cases we have seen only diarrhoea and weight loss. This is particularly worrying as it can lead to disease easily being missed. It can occur at any age, but typically occurs in growers and finishers.

Cases of unexplained diarrhoea and/or wasting must be immediately reported to your vet and investigated with laboratory testing for swine dysentery. If your vet is unsure, they must contact APHA to understand how to submit this. We really need to know where this disease is in the UK, so keepers should sign up to the significant diseases charter, through which they can anonymously report any cases.[18] Those signed up to the charter can then also see the cases that are occurring around them.

Post-Weaning Multisystemic Wasting Disease (PMWS)

PMWS is caused by a virus called porcine circovirus type 2 (PCV2). Piglets at six to eight weeks appear depressed, look very hairy and often jaundiced, and may have diarrhoea and lose weight. They will also be more susceptible to other diseases. Pigs may also show respiratory signs, for example struggling to breathe. There can be a high number of deaths, which peak at around nine weeks of age. At post-mortem, lymph nodes will be very large.

The disease can be vaccinated against, which is standard for commercial pigs in the UK due to the commonness of this disease. The situation on smallholdings is unknown and therefore most will only vaccinate after they come across the disease.

Later, pigs with PCV2 can experience another syndrome called porcine dermatitis nephritis syndrome (PDNS). PCV2 can also cause respiratory and reproductive signs.

Ileitis

Ileitis describes two conditions caused by the bacteria *Lawsonia intracellularis*. In recently weaned pigs the bacteria causes poor growth, often with diarrhoea. Diarrhoea can be bloody, and therefore swine dysentery must be tested for alongside this disease. In finishers, sudden death occurs and at post-mortem the guts are found to be filled with blood. The bacteria is spread from pig to pig via the ingestion of faeces and is often introduced into a herd via a carrier pig. Vaccines are available in an injectable form, which can be considered where disease is experienced.

Intestinal Parasitic Worms

There are several types of worms that can cause intestinal problems in pigs kept on UK

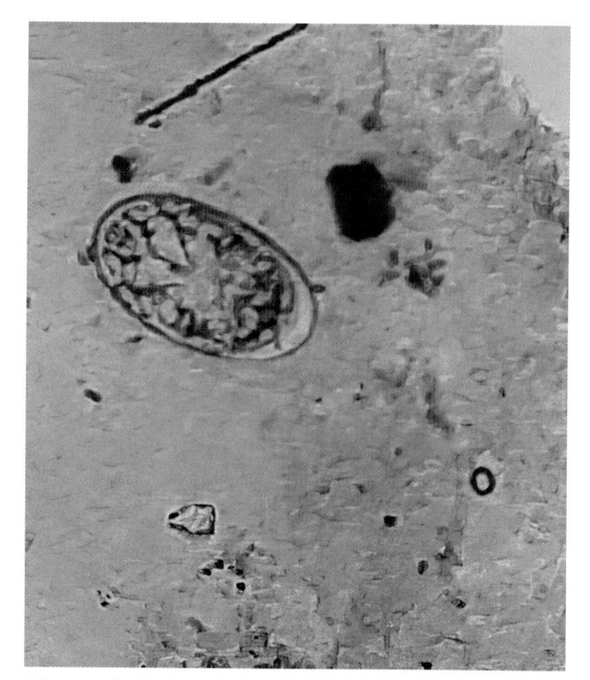

The typical shape of a strongyle egg under the microscope.

smallholdings. We will discuss their control in Chapter 12, 'Parasite Control'. The first category are the strongyle worms, which are grouped together because they have the same shaped egg.

The strongyle worms are:

- *Hyostrongylus rubidus*: The 'stomach worm', causing inflammation of the surface of the stomach, which can then cause a stomach ulcer.
- *Oesophagostomum dentatum*: Nicknamed the 'thin sow syndrome', causing loss of condition in sows due to inflammation of the large intestine. Immunity to this parasite through exposure doesn't develop.

Trichuris suis (whipworm) is very different from the strongyles. Disease is uncommon but important because the clinical signs look very similar to swine dysentery, and therefore must be investigated. Generally, infections do not cause noticeable disease, but in large numbers cause diarrhoea, sometimes with blood, in growing pigs.

Atresia Ani

Atresia ani is a condition of newborn piglets where a layer of skin prevents the rectum opening to the exterior. It is genetic to some degree but there are likely other factors at play. To check this in a newborn piglet, place a rectal thermometer very slightly into the rectum and check that no skin stops its path.

If left untreated, the pig generally can't defecate, meaning that it develops a pot-bellied, swollen abdomen and has to be euthanased. The condition must either be surgically repaired by a vet or the piglet euthanased as soon as it is confirmed. Keepers must not operate themselves as it will cause severe pain in the piglet and is illegal (as it is an act of veterinary surgery[1]).

Intestinal Torsion

Intestinal torsion is a common cause of sudden death in growing pigs where excessive gas forms in the gastrointestinal tract from recently eaten foodstuffs, leading to torsion of the small intestine. This might be due to pigs eating too much and therefore often affects the best pigs, and is especially common after feed has been withheld. It occurs acutely with pigs experiencing severe pain and showing a distended abdomen. Many are found dead, due to the twist blocking veins coming from the intestine, and therefore blood cannot return to the heart. No treatment is available and pigs should be euthanased where this is suspected. To prevent this occurring, feed pigs regularly so they don't become starved and gorge their feed, feed non-fermented foodstuffs, and have ample trough space for all pigs at once.

Stomach Ulcers

Stomach ulcers in pigs can present as a persistent low-grade pain, or pigs will be found dead due to a perforated ulcer. Ulcers can present with weight loss and reduced appetite, an arched back and reluctance to move. Pigs may vomit, and sometimes digested blood can be seen in faeces, which has a dark, coffee-granule appearance. Sometimes pigs will be pale due to blood loss. In our experience, this condition is most seen on smallholdings in sows post farrowing. It can also be seen pre-farrowing or in rapidly growing pigs.

The condition is often multifactorial. Feeding finely ground feeds, especially those consisting of cereal or wheat, can irritate the stomach lining. This fine grinding is necessary to formulate diets into pellet forms, so pelleted feed is especially implicated. Stress will make an ulcer far more likely, for example the stress of giving birth or transport. A pig not eating can also cause a stomach ulcer, which is why sick pigs

Pig vomit is generally quite easy to discern from pig poo.

will often have a certain degree of stomach ulceration.

Often the bacteria *Helicobacter* is suspected to be implicated, but it is often thought to infect the ulcer after previous damage, and can be found in many normal stomachs too. We mention this because your vet may want to give an antibiotic in the very short term.

Overuse of non-steroidal anti-inflammatory drugs (NSAIDs) such as meloxicam can also cause stomach ulcers. Animals on long-term use of these drugs, due to a chronic condition such as arthritis, are at risk of developing stomach ulcers, along with renal failure. It is essential to dose according to your vet's instructions. If you have a pig on long-term NSAIDs that goes off its food and becomes depressed, you must speak to your vet so that appropriate action can be taken.

Animals that could be experiencing stomach ulcers need veterinary treatment and close veterinary supervision. Feeding an oat porridge (not prepared in the kitchen) with mashed banana can tempt them to eat, and flat coca cola can help the nausea.

To prevent the condition, stress should be minimised, and feed should not be withheld unless absolutely necessary.

Hernias

Hernias describe a piece of gut being somewhere other than the abdomen, and can be umbilical or scrotal. They are genetic to a degree, but umbilical hernias can also be enlarged by unhygienic conditions at farrowing.

They are variable in size, and some cause the gut to become strangulated while others cause no problem at all. This strangulation can limit growth or can lead to death. Where hernias are large (more than about 15cm) or if the skin is broken, a pig will be unfit for transport and should be humanely euthanased on farm rather than being sent to slaughter.

Hernias can recess, so it is important to ensure that you take note of any at birth and do not breed from these pigs and preferably their siblings, due to the genetic element.

A mild umbilical hernia. This pig will probably cope well until finishing weight, but it shouldn't be bred from.

A testicular hernia in a piglet. This doesn't seem to be restricting his growth currently, but may do later on.

Bacteria can enter the piglet through abrasions such as these.

LAMENESS AND LEG ISSUES

Because pigs are a prey species, they are incredibly stoic and resistant to showing pain. This does not mean that they are not feeling pain, but they cannot show it without risking being picked off by a predator. This is especially important when observing lame pigs, and it is imperative that pain management is sought early where pigs are lame as you can be assured that they are not showing the full extent of the pain they are feeling.

Joint Ill

Joint ill describes a condition where bacteria locate themselves in the joints of young piglets and cause hot, swollen and painful joints due to arthritis. Bacteria will have entered through another site, such as any abrasions or through the navel. They move into the blood and then to the joints. Where holdings clip teeth or dock tails this is more likely, however this should not be relevant for smallholdings. Causative bacteria can range from streptococci, staphylococci, *E. coli* and more.

Piglets present with swollen joints and lameness at any stage pre-weaning. If the bacteria also move elsewhere, piglets may die suddenly, for example due to meningitis. The disease can prevent them from competing for milk and as a result they can quickly deteriorate. The condition is very painful and piglets should be monitored daily to be sure they are walking normally, and veterinary intervention sought immediately if not.

Navel dipping in iodine at birth can help prevent this disease, as can good colostrum management. Good hygiene at farrowing is crucial, removing faeces promptly and thoroughly cleaning and disinfecting between litters.

Joint Infections in Older Pigs

Older pigs are also commonly subject to bacterial joint infections that cause arthritis, but these have different causes in terms of triggering factors and implicated pathogens. The condition is seen most in growing pigs, and the bacterial cause can range from *Mycoplasma hyosynoviae, Haemophilus parasuis*, erysipelas, streptococci, and more. Each has a slightly different presentation, and they can be difficult to discern.

We can see lameness in one or more limbs and/or swollen and hot joints in growing pigs. Many of these bacteria cause problems after stress, particularly transport. The condition is painful and requires immediate veterinary attention. Joint infection can be difficult to differentiate from non-infectious arthritis, and there are very different treatments required for different bacterial causes. Inconveniently, erysipelas and streptococcus have opposite treatments to mycoplasmal lameness. For erysipelas and streptococcus we generally use a penicillin antibiotic, which would not work for mycoplasma as mycoplasma do not possess the cell-wall structure that penicillin disrupts. It is for this reason that achieving a diagnosis by involving your vet is so important!

Non-Infectious Arthritis

Fast-growing pigs can be predisposed to a cartilage issue, meaning that they can experience arthritis at a much younger age than we would commonly expect, even whilst still growing. This can be a particular problem where commercial pig breeds are kept as pets, as they have not been bred for the purpose of a long life.

Pigs with poor conformation also tend to develop osteoarthritis very frequently. Kunekune limbs can point outwards, leading to structural issues that will eventually cause arthritis. Coupled with obesity that develops in many pigs over time, this can present a real problem, and means it is imperative that pigs are not overfed and have space to exercise.

Pigs may be obviously lame on one limb or can present as reluctant to move, with a stiff gait, and have an arched back. Once arthritis develops, pain must be managed under veterinary supervision. Management techniques such as ensuring that pigs are on a non-slip surface can be helpful.

Fractures

A fracture occurs when a bone is broken, generally due to some kind of trauma. We often consider a fracture easy to diagnose, however pigs can be much more challenging due to having very large bodies of muscle. These act to stabilise fractures, meaning that the classic wrongly angled limbs and protruding bones are rarely seen. Coupled with a pig's tendency to not show pain, this can be really dangerous, and treatment can be accidentally withheld from pigs in excruciating pain.

Fractures of bone in the legs generally present as a lameness in one limb. The pig

This pig couldn't stand on her back legs and was dragging them along the floor. It is likely she had a spinal fracture or a slipped disk in her spine.

may still be putting weight on the limb but will often hold it in the air when standing still. It may have a 'swinging in the breeze' appearance or be held at an angle.

Spinal fractures often present with a pig that cannot stand on its back legs, which are flaccid, and the pig walks dragging them along the floor. This presentation can also be caused by a spinal abscess, or really anything that occludes the spinal cord, but the prognosis for each is poor and the welfare state of the pig must be seriously considered.

Pigs, adult boars especially, can experience a fracture of their femoral head, which is the top of the bone that connects into their hip joint. This often happens while they are serving. They will be severely lame on a back leg, but are nevertheless generally putting some weight on to it. The gluteal muscle collapses in a fairly classic presentation.

Fractures are a veterinary emergency, with euthanasia as the likely outcome. The shape of pigs' limbs do not lend themselves to casting, and the length of time that self-healing would take is unacceptable in terms of welfare. Where there are multiple fractures on a holding, a nutritional cause must be considered. Pigs commonly develop issues with calcium, phosphorus or vitamin D, such as rickets, where nutrition is substandard.

A pig holding its leg in the air whilst standing raises alarm bells for a fracture. From the 'dropped' appearance of the gluteal muscle, this looks like a fracture of the femoral head.

Foot Problems

Foot problems range from simple claw overgrowth, for which foot trimming will be necessary (*see* Chapter 15, 'Common Procedures'), or more serious abscesses. Abscesses occur after bacteria have entered into the foot and it will be very painful to walk on, swollen and hot. Eventually the abscess will burst and the skin will die. This intensely painful condition will likely require veterinary intervention, possibly under sedation.

Foot issues such as this abscess need veterinary attention.

RESPIRATORY DISEASES

Respiratory disease pathogens in pigs generally work together. The initial disease is often sparked by a stress event, for example transport, making pigs more likely to succumb to a virus, or to enzootic pneumonia, which is caused by a bacteria that behaves quite similarly to a virus. These cause damage at the top of the respiratory tract. You will hear coughing and sneezing, similar to us getting a cold or flu, and damage to the respiratory defences can occur.

If the conditions are right, secondary pathogens (generally bacteria) can then invade deeper into the lungs and cause pneumonia. Here we will see more serious signs of pigs struggling to breathe, lying down and extending their neck to take in air, and sometimes dying.

Pig lungs are assessed at slaughter, and producers should pay really close attention to this feedback. Widespread respiratory disease is most associated with a large number of animals being in close proximity to one another, and therefore smallholdings are generally not a hotbed for respiratory disease.

Ascaris Suum, Milk Spot Liver

Ascaris suum is a parasitic worm that is incredibly common, especially on smallholdings. These worms are huge as adults (up to 40cm long) and the eggs are very resistant in the environment. The eggs are carried by birds and rodents so can enter a holding at any point.

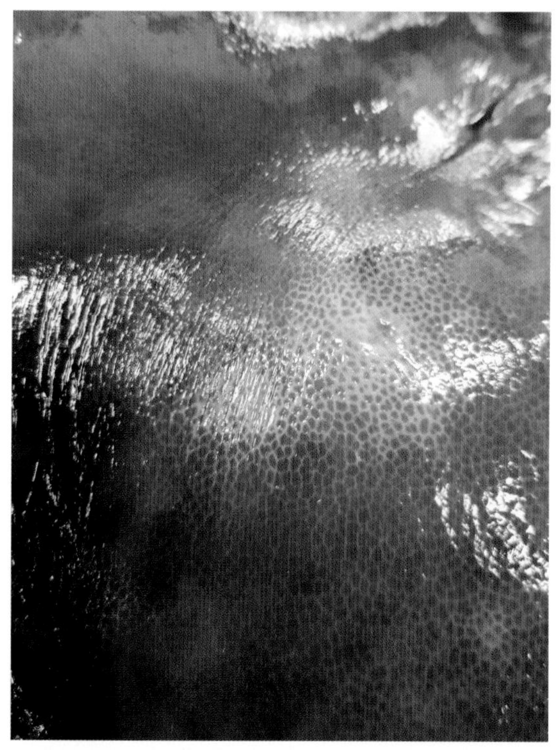

A classic example of a milk-spot liver at post-mortem.

The life-cycle occurs as follows:[25]

- Eggs containing larvae are swallowed by pigs from pasture or the environment.
- The larvae then hatch and burrow into the gut. From this point they migrate through the liver and then the lungs. It is through this migration that the clinical signs of liver disease and coughing or 'thumping' may be heard. In addition, 'milk spot liver' may be observed at the abattoir, which will lead to condemnation.
- Larvae are then coughed up, swallowed, and finish up as adults in the gut, about three weeks after they were initially ingested. Here, blockage and death is possible, or huge worms can be passed in faeces and noticed by unsuspecting smallholders!
- Adults will then begin laying eggs that may be detected in the faeces. These take two weeks to mature on pasture, at which point the larvae will hatch when they are eaten by other pigs.

Besides the clinical signs above, pigs will experience slower weight gain, and the lung damage can exacerbate other respiratory conditions. Control is discussed in Chapter 12, 'Parasite Control'. This parasite is zoonotic and therefore it is imperative to practise good personal hygiene around pigs.

Porcine Respiratory and Reproductive Syndrome Virus (PRRS)

PRRS is one of the most important diseases economically in UK pigs,[26] but the prevalence amongst smallholding pigs is not known. As the name suggests, severe clinical signs are seen, which is dependent on the age of the pigs infected.

The life cycle of *Ascaris suum*.

Respiratory disease is observed in growing pigs and is often concurrent with other respiratory diseases such as enzootic pneumonia or influenza. Piglets from around five weeks of age will cough and sneeze, and growth will slow. PRRS enables secondary bacteria to invade and cause pneumonia, and weakens pig immunity, meaning that other infections are more likely. The disease can't be cured, but supportive care is very important, especially in terms of secondary infections.

Many herds have PRRS continuously circulating and therefore see quite mild clinical signs of coughing and reduced efficiency. This can be more severe after times of stress. Other herds will not have PRRS present, meaning that signs are particularly severe if it is introduced, as the pigs' immune system has never seen the disease.

The virus is mainly transmitted from pig to pig, but transmission through objects, vehicles and people is also important. The virus can also move short distances through the air from farm to farm. PRRS is therefore the perfect reason to operate good biosecurity.

Once in a herd, eradication is often impossible. Many positive herds will vaccinate for the disease, a subject that will be covered more fully in Chapter 11, 'Vaccinations'. General management is very important, especially stopping piglets becoming cold at weaning, and separating age groups.

PRRS is often nicknamed 'blue ear', referring to the blue-tinged ears that often develop in sows with the disease. Unfortunately this can lead to keepers believing that any pig with blue ears has PRRS, which is generally not the case as any sepsis (for example with erysipelas) can lead to blue ears as blood struggles to move round the body – so it is not a definitive sign!

Finally, it is important to note that two main strains of PRRS exist, a North American strain and a European strain. The European strain is far less serious in terms of clinical signs of disease than the North American strain, and does not spread nearly as far. We currently do not have the North

American strain in the UK, and we absolutely must keep it this way! Do be aware of this difference if you are doing any research into PRRS, and be sure that you are reading literature on the European strain if you are wanting to understand the UK situation.

Swine Influenza

Much like influenza in humans, flu viruses tend to infect large numbers of pigs very quickly, but there are few deaths. Affected pigs will have a persistent, dry cough. They will appear depressed, will stop eating, and may sneeze and have discharges around the eyes and nose. As long as they aren't infected by another pathogen, they generally recover after a few days, but pregnant sows will commonly abort during this disease. The disease is most common in the winter. As with people, there is no effective treatment, just supportive therapies such as giving pain relief when needed, and ensuring plenty of access to fresh water.

Interestingly, pigs can be infected by influenza viruses from both people and birds. We then worry about a whole new strain of influenza 'mixing' inside a pig, which could cause our next global pandemic! Therefore it is really important that people with flu do not go near pigs.

Porcine Circovirus Type 2

As described under post-weaning multisystemic wasting syndrome, PCV2 is also often isolated from the lungs when respiratory disease is present and can cause coughing.

Enzootic Pneumonia (*Mycoplasma hyopneumoniae*)

Enzootic pneumonia (EP) is caused by the bacteria *Mycoplasma hyopneumoniae*. This bacteria is very, very small, so behaves like a virus in many ways. It is incredibly common and generally causes low-level and persistent coughing and reduced growth rates. Producers should pay close attention to abattoir feedback, as EP causes some quite classic lung signs that will be reported.

This lung tip has the typical appearance of consolidation from EP, though it is likely that some other bacteria are also involved.

Commercially, herds will either vaccinate for this disease or they will have eradicated it, and will test regularly to prove they are free from it. Many smallholders also vaccinate for this disease, especially those taking pigs to show. The disease spreads through direct contact or from pigs sharing the same airspace, so infection at shows is likely.

Actinobacillus Pleuroneumoniae (APP)

APP is a bacteria that causes catastrophic clinical signs, very suddenly in growing pigs (post-weaning). It doesn't need viruses to spark infection, but the disease can be more severe where other pathogens are present. Pigs will struggle to breathe, appear dull, have a high fever (40.5°C), and stop eating.

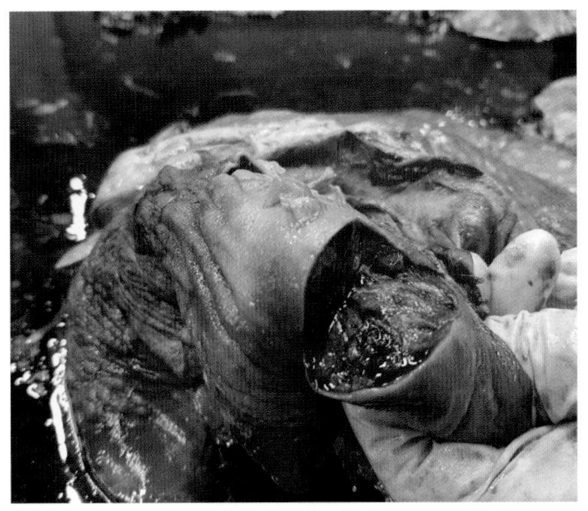

Hopefully it can be seen how this lung, infected with APP, would not have been able to function.

Some may suddenly die. A less acute form can be less severe. There is an effective vaccination against this disease.

Secondary Bacterial Pneumonias

Following initial infection from one of the above diseases, bacteria can invade to cause a secondary pneumonia. These can include pasteurella, Glässer's disease, streptococci and many more. The key here is providing the optimal conditions so that primary causes don't lead to disease of the upper respiratory tract, and supporting pigs with upper respiratory tract disease so that bacteria don't then cause pneumonia. Pigs with pneumonia are very, very sick and require 'intensive care' facilities, with close veterinary supervision and monitoring.

Atrophic Rhinitis

Atrophic rhinitis is a disease of young (three- to eight-week old) pigs caused by two bacteria that often work together: toxigenic *Pasteurella multocida* and *Bordetella bronchiseptica*. It is now virtually eradicated in commercial pig herds due to vaccinations and management interventions; however, it still crops up on smallholdings. The disease is characterised by a shortened and twisted snout, and sneezing. Pig growth will also be affected.

A pig with atrophic rhinitis, which has caused the slanting of his snout.

The disease is unpleasant, and not one that we want in the pig sector, so call your vet and send them photos if you see these signs. Vaccination is available and should be practised in positive herds.

LIVER DISEASE

Clostridium Novyi

Clostridium novyi causes fatal and very acute liver disease, generally in sows. It may appear secondary to other diseases, such as ascarid parasitic worms, which set up the necessary conditions in the liver for it to thrive. The condition is tricky to diagnose because the effect on the liver and carcass is the same as the natural progression after death (especially in the warm weather in which this condition tends to occur), so a post-mortem

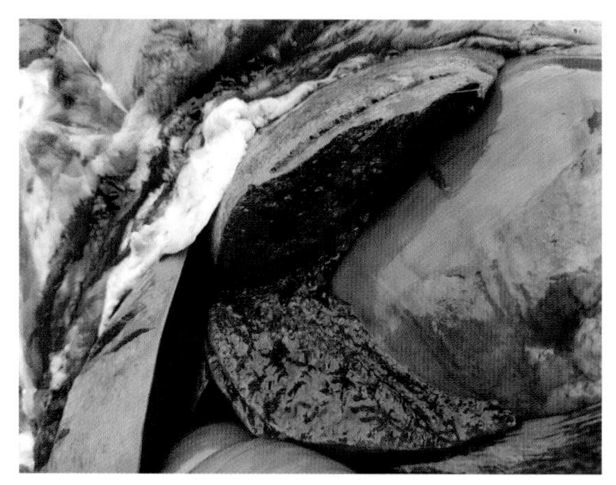

The classic appearance of a liver at post-mortem from a sow that died of *Clostridium novyi*.

needs to be carried out very soon after death to be helpful. Always speak to your vet in the case of sudden death to rule out notifiable disease. Clostridia will be high on their list in cases of sow death where the carcass very quickly becomes purple-green, gassy and puffy. Clostridia bacteria are commonly found in soil where they live for many years, so outdoor production holds specific risk for these diseases.

Coal Tar Poisoning

Coal tar poisoning is regularly seen on smallholdings, likely sources being clay-pigeon fragments, roof asphalt and waste from road tar.[20] Pigs appear dull, stop eating and die. Paddocks should be regularly walked to monitor the presence of toxic substances.

DISEASES AND CONDITIONS OF THE SKIN

Greasy Pig Disease

Greasy pig disease is an excellent description for this condition, which is caused by the bacteria *Staphylococcus hyicus*. Piglets (one to five weeks old) look and feel greasy, but won't scratch. They look grey, hairy, and have scabs all over their body. The bacteria can also cause the ear tips to die off later in life.

Sores like this on the tips of the ears are often also caused by *Staphylococcus hyicus*.

Pigs with the disease really struggle and often become dehydrated and die so need urgent veterinary treatment. The disease generally results from skin abrasions that have become infected, such as due to rough bedding.

Porcine Dermatitis Nephropathy Syndrome (PDNS)

PDNS is the second syndrome caused by the virus PCV2. Twelve- to fifteen-week-old pigs develop red-purple blotchy patches or spots over their body. They stop eating, appear dull, and may have a stiff walk. The skin disease is accompanied by severe kidney disease, meaning that pigs are very sick and can die.

As previously mentioned, this disease is clinically indistinguishable from the swine fevers, particularly CSF. Therefore by law, you need to call a vet immediately if you see this. Where disease is present on a holding, piglets should be vaccinated.

External Parasites

We will discuss control and prevention of these parasites in Chapter 12, 'Parasite Control'.

Scabies

Scabies in pigs is caused by a pig-specific sarcoptic mange. When we refer to a pig as suffering from 'mites', we are generally talking about this parasite. Pig scabies can transiently bite people but won't live on our skin.

Scabies is of huge importance in UK pigs and causes scratching, ear crusts, thickened skin, abrasions, scabs and scurf. Affected pigs can shake their head, sometimes leading to swollen ears. More marked reactions, such as papules and redness over the body, can be seen in pigs experiencing a hypersensitivity reaction. The disease tends to get worse during the winter as sunlight reduces the viability of the mites.

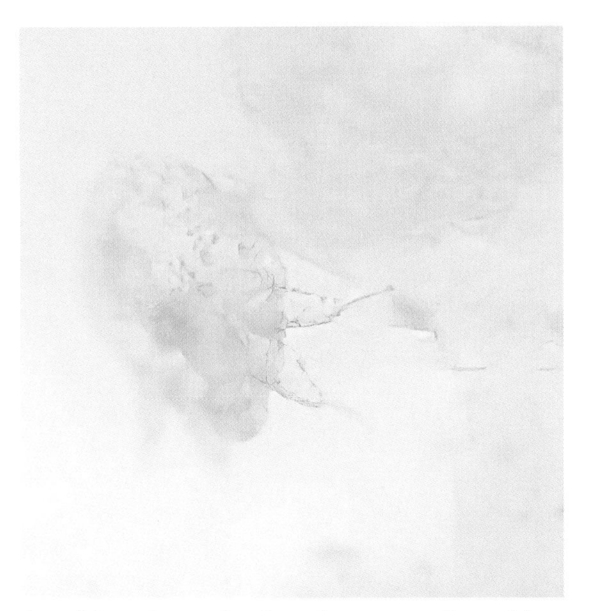

A scabies mite under the microscope after taking some ear crusts from a pig.

The whole life cycle takes place on the pig in ten to twenty-five days (however, environmental contamination does still occur, making control difficult). Mites generally only survive off the host for about three days,[27] so try always to leave housing empty for one week before moving other pigs in. Mites are spread directly between pigs.

The mites are rarely visible on the pig as they are small, and therefore we look at ear-crust scrapings under the microscope to diagnose this parasite. With this method, we don't consistently find mites even when they are present, so most vets will suggest treatment when excessive itching is observed.

Lice

Pig lice are entirely pig specific and have a life cycle of about twenty-five days, which takes place wholly on the pig. Like mange, it can only survive for two to three days off the pig. Lice also cause a pig to scratch excessively, and can damage the skin.

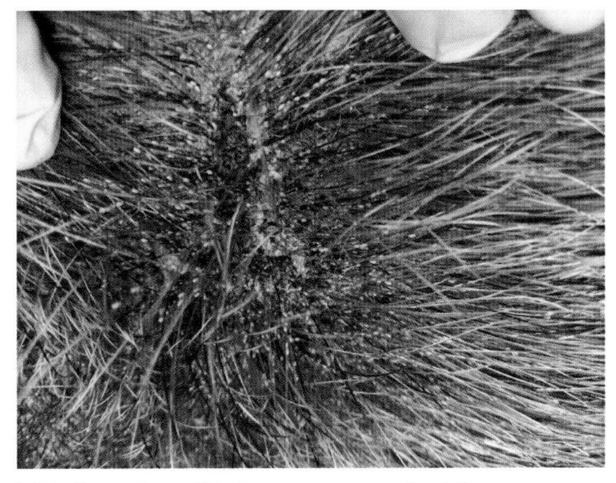

This boar is suffering a very marked lice infection - lice and scurf can be seen close to the skin.

Pigs love to scratch on things, and this is perfectly normal behaviour. However, too much scratching should raise suspicions of external parasites and prompt a closer inspection.

The moult on this Kunekune can be seen spreading from the top of his back down his flanks.

Lice can be seen as nits, which are the lice eggs. Nits are white and are found on hair shafts very close to the body. These occur mostly on the head, neck, flanks and inside of the legs. Lice can cause anaemia of young piglets, so control is particularly important in these.

Kunekune Spring Moult

The Kunekune spring moult is a normal condition that happens to Kunekunes, or pigs with some Kunekune in them, during spring to summer. The pig will be completely well and will lose hair from the top of its back, extending down the flanks. They will not be itchy.

Sunburn

The lack of hair on most pigs means that skin can be more exposed to the sun, and if not gradually introduced, pigs may burn. Pale-skinned, less hairy pigs are more prone to this. Sunburn can also lead to abortion. Pigs with sunburn will have reddened skin, will show discomfort, and may be reluctant to move. Breeding sows may also be reluctant for a boar to mate her. Blisters may result in extreme cases.

If sunburn occurs, pain relief is essential, so please phone your vet for advice regarding this. If an individual pig is prone to sunburn, it may be worth investigating health problems such as liver dysfunction or toxic plants that the pig may have eaten, such as parsnips or parsley.[20]

Refer to Chapter 4, 'Pig Housing and Environment', for practical steps on how to prevent this condition.

This pig has caught the sun over the pink skin on his back.

DISEASES OF REPRODUCTION

Parvovirus

Parvovirus is a virus that causes reproductive problems such as stillbirths, mummified piglets, embryonic death and infertility. Together we call these things SMEDI. It is very common and is therefore the reason that we suggest parvovirus as a core vaccine for breeding pigs. Parvovirus also survives for a long time in the environment.

Unless a pig is pregnant, parvovirus sits in the gut not causing a problem, possibly infecting other animals through faecal shedding. If a pregnant pig is infected, the signs that are seen will depend on the stage of pregnancy at the time of infection. Infection before day thirty of pregnancy results in the embryo being lost. Infection between thirty and seventy days can result in foetal death and mummification. Some foetuses survive and are born alive, but are persistently infected with the virus. Those infected after day seventy of pregnancy mount an immune response, clear the virus, and are healthy at birth.

A classic presentation at birth is a litter that is totally dead with a spectrum of normal-looking piglets right through to very small and mummified foetuses. The virus moves through the litter slowly, causing this presentation. Generally, the sow will farrow at the standard 115 days. Parvovirus cannot be treated and should instead be prevented with vaccination. We will cover this in Chapter 11, 'Vaccinations'.

Porcine Respiratory and Reproductive Syndrome Virus (PRRS)

PRRS also causes reproductive problems in pregnant sows, and disease is dependent on the time of infection. Early infection in the pregnancy causes the embryo to die, and the sow will come back on heat. From about thirty to eighty days into gestation, infection will result in mummified foetuses, but unlike parvovirus all the mummified piglets will be the same size. Later than this causes stillborn piglets. Sows are likely to abort (give birth to dead piglets earlier than 109 days into pregnancy) or farrow early (before 113 days into gestation).

Piglets born alive will be weak and may be anaemic and have diarrhoea. Sows may also become sick and may show the classic 'blue ear' presentation. Boars will infect sows with PRRS virus through their semen, which is a specific risk when hiring boars. Artificial insemination centres are tested continuously to ensure that semen does not contain PRRS.

Control is through prevention of the disease entering a holding by practising good biosecurity. Vaccination can be instigated, but please go to Chapter 11, 'Vaccinations' to find out why this may not always be sensible.

Leptospirosis

There are many species of leptospira bacteria, some of which are associated with pig reproductive disease. Clinical signs range from early embryonic loss, meaning that a sow comes back into heat, discharges from the vagina containing pus, abortions, weak piglets and sometimes stillbirths and mummified piglets. Rodents can be significant in terms of disease spread.

Leptospirosis can be zoonotic and therefore you must always wear gloves when touching aborted material or stillbirths. Vaccines are now available in the UK combined with erysipelas and parvovirus.

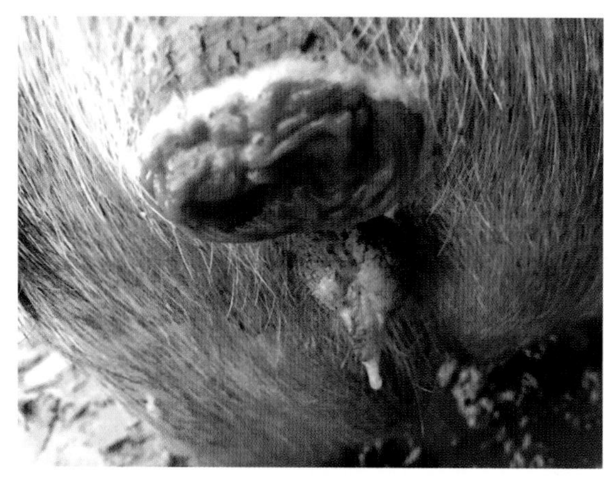

A mild rectal prolapse that will likely pop back in quite quickly of its own accord.

Other Infectious Causes of Reproductive Loss

Any disease that causes a sow to experience a fever is likely to cause her to abort. Key examples are erysipelas and swine influenza, which have been covered elsewhere. PCV2 can cause late-term abortion and stillbirths.

Diseases of the Sow Shortly after Giving Birth

Prolapse

Occasionally pigs can eject an organ that should be held in the body.

Rectal prolapses can occur in growing pigs. They will generally go back in naturally, but pigs must be separated into a hospital pen so that other pigs don't attack it, and be given pain relief under veterinary supervision. Severe cases should be euthanased, and old-fashioned 'tricks' using bands or pipes must not be practised.

A vaginal prolapse in a sow can often be replaced with veterinary assistance, however it is very unlikely that a whole uterine prolapse can be replaced. Generally a sow is dead from shock before a vet can even arrive, and if not it is generally kinder to euthanase the sow. This will then leave a really difficult situation with piglets. Ideally they would be fostered by other sows, or they may need to be hand reared, and in the worst case scenario euthanasia of the litter may be most appropriate to prevent further suffering.

Postpartum Dysgalactia Syndrome (PDS)

This syndrome covers a range of conditions, also known as farrowing fever, toxaemic agalactia lactation fever, and mastitis-metritis-agalactia. The latter has been used frequently but has led to confusion that this disease is caused by an infection (which it is not), so we will refer to PDS.

The disease is generally seen in sows twelve to seventy-two hours post farrowing, and presents as a sow that has hungry piglets due to a reduction or cessation of milk (dysgalactia/ agalactia). The udder will be hard, and there may be oedema (fluid congestion) of the udder, hence a label of mastitis. Other signs such as a vaginal discharge may also be present (hence a label of metritis). Sows may stop eating and/or be constipated. Hungry piglets may be noisy at first, followed by diarrhoea and weight loss. If action is not taken quickly, piglets can soon starve to death.

The vast majority of cases are due to management issues such as inadequate nutrition, heat stress or inadequate water intake. Fat sows are at particular risk for the condition, as are sows with small litters, or litters not drinking due to illness.

There are two mechanisms that may lead to this condition:[28]

- A sow needs to be stimulated continuously to produce milk by piglets sucking on all her teats. If this does not occur or if it is inadequate, milk will not be let down and affected teats will 'shut off'.
- Where milk is produced, this needs to be drunk by piglets quickly so that pressure in the udder doesn't build up. If allowed to build, this pressure damages cells, and this damage can then lead to infection with bacteria such as *E. coli*; however, it is important to consider this secondary to the build-up of pressure.

Veterinary attention is crucial to ensure that the condition is properly differentiated from the bacterial causes of mastitis described below. Pain relief and anti-inflammatories as well as oxytocin (to promote milk let-down) can be administered under supervision from your vet. Piglets will require supplementary feeding.

Due to several possible causes, this condition can be difficult to prevent, but veterinary attention should be sought to understand the specific risk factors for your holding. Bran supplementation to prevent constipation around farrowing can be helpful, as well as practising good nutrition, as described in Chapter 5, 'Feeding'. High water intake around farrowing must be encouraged, for example by soaking feed and feeding fresh food often during lactation. Sows must not be fat at farrowing, and housing should decrease the risk of a sow becoming too hot, whilst stopping piglets from chilling.

Acute Mastitis

Sometimes an infection that is introduced into the udder can be a primary cause of mastitis - for example a bacteria such as *E. coli* from sharp piglet teeth. Here the udder will be very hard, hot and painful. The sow will appear very sick and will have a very high temperature (above 40°C). Her milk will be abnormal - it will have clots, be watery or be bloody. Veterinary attention must be sought immediately. Unhygienic conditions are often to blame for this condition.

Metritis

Metritis (inflammation of the uterus) can also exist post-farrowing due to bacterial infection. This is most likely if a sow does not pass the placenta, if a piglet is still stuck in the birth canal, or if there has been unhygienic manual assistance. Some discharge in the first few days after farrowing is normal as the uterine horns contract. However, discharge with a sick sow should be acted upon, as should discharges that are pus-contaminated, of high volume, or that

This discharge was from a sow that was poorly post farrowing. This set of clinical signs would raise alarm bells for metritis.

continue for several days after giving birth. The infection may ascend into the bladder and kidneys, where it can lead to death due to sepsis.

If manually assisting a sow, scrupulous hygiene must be practised. The placenta must also be laid out after farrowing to ensure that the entirety of both horns has been ejected.

Urinary Tract Infections

Urinary tract infections can move up to the kidney and cause death, generally in sows. Many different bacteria can be isolated, but the condition can be caused by inadequate water intake, or infection is sometimes introduced at breeding. Pigs may urinate frequently with excessive straining. Urine may be bloody or cloudy, and crystals may be seen with a microscope. Pigs will become sick very quickly, or alternatively can just be found dead. Prompt veterinary treatment is imperative for this condition.

CONDITIONS MOST COMMONLY SEEN IN PET PIGS

Blocked Urethra

Male pigs (especially castrated males) can experience a blocked urethra. Bladder stones form which become trapped in the urethra, meaning the pig cannot urinate freely. The condition will cause pigs to strain to urinate, often looking as if they are constipated. This excessive straining can lead to a rectal prolapse. As well as the sex of the pig, predisposing factors for bladder stones include an inappropriate diet, interruption to water intake (such as from frozen pipes or bullying), and urinary tract infections.[29, 30]

Urgent veterinary treatment is required for these cases, and pigs will require constant availability of fresh water. Prognosis can be poor, however - in our experience a combination of veterinary medicines, supportive care and alternative remedies, such as cranberry juice and uva ursi, can be successful.

Cancers

Pigs, like all species, can develop cancer, but this is rare, and as with people, is more commonly found in old pigs. Lymphoma is thought to be the most common cancer in pigs.[31] A pig with lymphoma could present with enlarged lymph nodes and weight loss. If the gastrointestinal system is affected, diarrhoea can also be seen.

Abdominal tumours in the liver and spleen are seen in older pigs,[32] particularly Vietnamese pot-bellied pigs, and these also present with a pig not eating, losing weight, and with an even bigger and more pendulous abdomen than is usual for the breed. Pigs with splenic tumours can suddenly die from internal bleeding if the tumour ruptures.

Hereditary tumours in pigs are again rare, but melanomas in miniature pigs can occur, which can spontaneously regress over time or can be more serious.[33]

Older pot-bellied and miniature pigs can develop uterine tumours. Clinical signs are observed including behaviour changes, abnormal vaginal discharge (often intermittent bleeding from the vulva), lack of appetite and a swollen abdomen.[34] The condition can be prevented by spaying young female pigs.

This little piglet's lump is a hereditary melanoma. Two of the piglets developed these tumours after a mating between brother and sister.

CHAPTER 10

MEDICINES

Sometimes smallholders are left with veterinary medicines to use as stipulated by their vet; they must therefore understand the rules and regulations around their use.

RULES AND REGULATIONS AROUND VETERINARY MEDICINES

Prescription-Only Medicines

Veterinary medicines are assigned categories according to their risk and the level of advice that therefore needs to accompany them. If a medicine is marked 'prescription-only medicine' (POM), this means it must be prescribed by someone who understands the animal(s) that the medicine should be used for, the condition they are experiencing, and who can certify that the medicine is appropriate in this case.

If a product is marked POM-V, this means that the prescription must be from a vet. This is because the product has been identified as having a narrow safety margin and therefore requires specialist knowledge. All antibiotics are assigned POM-V.

Prescription of a POM-V product must be following a clinical assessment of the animals under a vet's care, to lower the risk of a condition being misdiagnosed and animals suffering as a result. Occasionally a vet may deem it suitable to prescribe without a clinical examination, but to do this they must have seen the animals recently enough to have personal knowledge of the animals and their condition, enabling them to make a diagnosis and prescribe. Once prescribed, if a client would like to buy the medicine from another licensed supplier, the vet can provide a written prescription to enable this.

If a product is marked POM-VPS, this means that the prescription can be from a vet, a pharmacist, or a suitably qualified person who has been trained to have extensive knowledge about the product and how it should be used. Vaccines are often examples of this.

To dispense a prescription-only veterinary medicine, the supplier must be sure that the person administering the medicine is competent to do so safely, that they will only use it as authorised, and that they are aware of all warnings and contra-indications associated with the product. Repeat prescriptions have to be vet authorised, so please don't see veterinary practices as trying to stand in your way of treating a pig. They are most often just trying to fulfil their due diligence around the use of veterinary medicines, and to ensure that accurate and prompt diagnoses are being achieved where possible.

When a keeper is given a prescription-only medicine, by law they must only use that medicine in the way that it has been strictly prescribed. This means that you cannot use the medicine for an animal that it was not prescribed for, even if the condition looks identical to you. It also means that you cannot use the medicine for any conditions other than those for which it has been prescribed. Finally, this means that veterinary medicines must not be shared between holdings.

Before administering a medicine you must ensure that you are completely clear as to what condition is being treated, and how the medicine should be used. If you are unsure you must speak to the prescriber. If your vet decides that some medicines should be held on the holding to use in specific situations after a prescription, information pertaining to their use is best held in a veterinary health plan.

Licensing of Veterinary Medicines in Pigs

In order to be authorised for use in the UK, each veterinary medicine must be licensed for a species and for a condition in that species for which it has been proven safe and effective. To gain a licence, the medicine must be proved to be safe to use, be of good quality, and be efficacious.

In order for a veterinary medicine to be used in a pig, it must have been tested to understand the level that the medicine is safe to be ingested by a human. Without that, if the animal were to end up in the food chain, we cannot know the human health issues that a previous administration of that medicine may cause. This means that there are many medicines used in dogs, cats and humans that we are not allowed to use in any pigs. This includes pet pigs, even if the owner is absolutely sure that the pig will never end up in the food chain. There is currently no legislation available to do otherwise.

Occasionally we will want to use a medicine outside its licence. For example we may want to use it at a different dose, for a different length of time, or for a condition that it is not licensed for. We might also want to use a medicine in a pig that has only been licensed in a cow. As vets, we have a process that enables us to do this under very strict terms. One of those is that the withdrawal period must be at least twenty-eight days for meat. This means that you must not use a veterinary product outside the way that it has been prescribed without discussing this with your vet. Doing anything other than this is illegal.

Observing a Withdrawal Period

A withdrawal period is the length of time that must pass between the final administration of the medicine, and the point when the animal can be slaughtered to enter the food chain. It is very important to be sure that this won't cause problems before treating a pig with a medicine.

Necessary treatment cannot be withheld from an animal, therefore if a pig can't be treated with a desired medicine you and your vet must choose between the following courses of action:

- Delay sending a pig to slaughter until the withdrawal period is over
- Find an alternative product that will be as effective
- Humanely euthanase the pig

The rules around recording the withdrawal period and individually identifying a pig during the withdrawal period apply to all pigs, even pet pigs. The rules stipulate that the pig should be individually identified in the way that is described in the medicines records for the duration of the withdrawal period. If temporary identification marks (spray or crayons) are being used, these will probably need reapplying.

Please be aware that withdrawal periods are extended for pigs that are produced on organic holdings. These rules will be stipulated in the scheme standards of organic certification bodies.

UNDERSTANDING WHAT YOU HAVE BEEN PRESCRIBED

Before using a medicine, you must read the available information to ensure that you understand the safety of the product and any contra-indications to using it. This information is held in the Summary Product Characteristics (SPC), which should be enclosed in the packaging. If it is not, you can find all of these in the Veterinary Medicines Directorate's Product Information Database.[1]

HANDLING MEDICINES

Only handle medicines that you do not have an allergy to. If you have any known allergies, be sure to let your vet know when they are prescribing medicines for your pigs. Always wear gloves when handling medicines, and ensure that you are adhering to the information in the SPC.

Storage of Medicines

Most smallholders will find that keeping bottles of medicine in stock leads to them going out of date and/ or being used incorrectly. Therefore in the majority of cases it will be most appropriate to pick up single courses from your vet.

Store medicines in the way described in the SPC (Summary Product Characteristics: *see* box). Products may have temperature or light stipulations. If you find that a medicine has not been stored in the way specified, you will need to contact the manufacturer and inform them how long the medicines have been stored outside their specification, and they will be able to advise you.

For products that should be kept refrigerated, you must ensure that 2–8°C

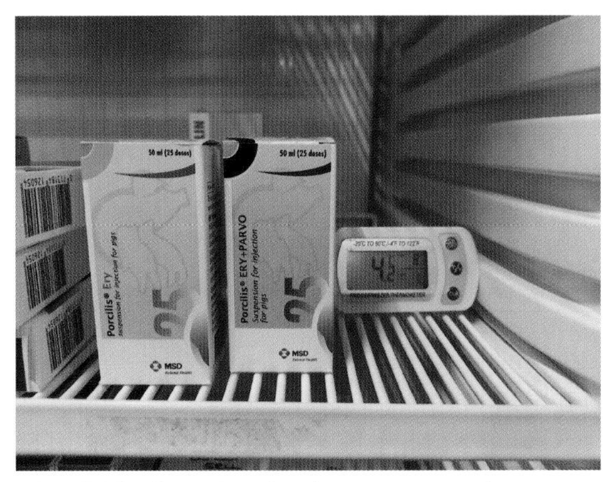

This fridge is maintained at an appropriate temperature.

will be maintained consistently. Therefore monitor a maximum and minimum thermometer in the days running up to collecting the medicine to verify that the refrigerator is suitable.

The cold-chain needs to be maintained consistently, even during transport. Be sure to bring a cool box or bag with you to collect products such as vaccines. Ice packs should be used, but it is really important that these products don't freeze, so wrap the ice packs in a towel and don't press the medicine against them.

Once collected, you need to be able to demonstrate that they have been stored appropriately, so records should be kept from readings of a maximum and minimum thermometer that you reset and record each day. Store medicines in an area where temperature fluctuations will be minimised, which can lead to products decaying more quickly. Try not to open the door as this will lead to temperature fluctuations. Don't overstock fridges as that can alter the temperature inside.

You should have a spill kit on hand where medicines are held. The storage area should be locked and only accessible to authorised persons. Food and drink should not be stored with medicines, including in a fridge.

Only use medicines that are in date, and check the SPC for how long the medicine can be kept once the bottle has been broached. This is generally twenty-eight days for injectable medicines, but will be much shorter for vaccines, which we will cover in more detail in Chapter 11, 'Vaccinations'. Some medicines will change colour when they go off (you might see black spots at the bottom of a bottle for example). Do not use these, and dispose of them properly. Some medicines will separate with storage so will need a really good shake before injecting.

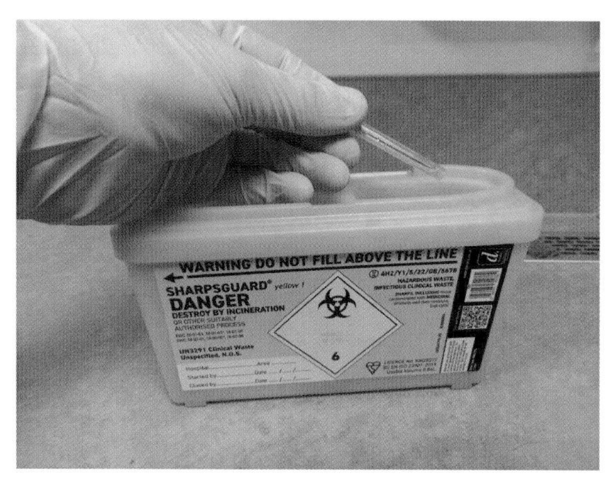

Needles should always be disposed of into a sharps bin.

Buying and Disposing of Medicines, Packaging and Sharps

Records on buying and disposing of medicines must be kept for a minimum of five years, even after the life of the pig. Keepers of pigs must keep proof of purchase and disposal of all veterinary medicines. For purchase, keepers must record:

- the name and batch number of the medicine
- the date of purchase
- the quantity acquired
- the name and address of the supplier

For disposal, the keeper must record:

- the date of disposal
- the quantity/amount of the product disposed of
- how and where it was disposed of

Any sharps should be disposed of into a sharps bin, which can be provided by your vet. They will then take this back when it is full and dispose of it for you.

Any medicine or medicine packaging must be disposed of as described in the SPC. This is really important to prevent residues of medicines causing problems to our environment. For example ivermectin wormers are toxic to aquatic organisms. They will generally need to be incinerated, so your vet will also be able to take these to dispose of (for a fee) with their own medicinal waste.

Recording of Medicine Uses

Medicines records must be kept for every medicine use for a minimum of five years. They should include:

- the name of the product
- the date of administration
- the withdrawal period
- the ID of the animals treated

If the medicine was given by a vet, the record must also have their name. The vet is legally required to provide you with the information that you need to complete your medicines records, or to complete it themselves.

It is best practice also to record the reason for giving the medicine, as this will allow you to understand the frequency of diseases amongst the pigs, although this is not required by law.

These records will be important for you to be able to fill out your Food Chain Information and eAML2 form, if you are sending pigs to the abattoir.

Sue and Stephen's medicine book, correctly filled in.

CLASSES OF VETERINARY MEDICINES

Veterinary medicines come in different classes, and these relate to the job that they do. We will go through the main classes of medicine that are used in pigs, which are pain relief and anti-inflammatory medicines, antibiotics, anti-parasitics and hormones.

Pain Relief and Anti-Inflammatory Medicines

To control pain and inflammation in pigs, two main classes of medicine are used: non-steroidal anti-inflammatory drugs (or NSAIDs) reduce both pain and inflammation; steroids just reduce inflammation.

NSAIDs

Common examples of NSAIDs used in pigs include meloxicam, ketoprofen and flunixin. These medicines are quite similar to ibuprofen for humans (noting that ibuprofen must not be used in pigs). They lead to the following intended effects:

- pain relief
- reduced inflammation
- lowered body temperature and fever

They are therefore brilliant tools to manage disease in pigs, both in terms of making the pig feel better, and reducing some of the effects of inflammation, which can be so harmful during disease processes. Generally, the more quickly a pig can be up and about after experiencing disease the better, and these medicines are fantastic to promote this. However, NSAIDs also have some challenging side effects that can happen in some cases:

- kidney disease if blood pressure is low
- stomach ulcers
- liver disease

Each will do this in slightly different ways, with different durations of activity and with slightly different side-effect profiles. For this reason they are licensed for different conditions and uses, and some are available in both injectable and oral formulations. It is really important that these are used only for the conditions that your vet has prescribed, and in the way your vet has prescribed, and not just for 'pain' that you may see on the holding. Firstly, you may find yourself inadvertently using these outside their licence, and therefore illegally; and secondly, you may also find yourself doing harm due to side effects. To give an example, a pig with a stomach ulcer will be in intense pain, but giving an NSAID is likely to make the stomach ulcer worse (and potentially rupture) due to its effects on stomach acid secretion. It is also really important that accurate weights are understood so that these medicines are not overdosed.

Steroids

Similarly, steroids can be very useful to decrease inflammation, especially where NSAIDs are inappropriate. However, these also have some challenging side effects:

- increased thirst and increased urination, so pigs on steroids must have unlimited access to fresh water
- increased appetite and weight gain
- hormonal conditions such as diabetes and Cushing's disease
- behavioural changes

It is also important to note that NSAIDs and steroids are not licensed to be given together.

Antibiotics

Antibiotics (and antibacterials) are substances that kill or inhibit other microorganisms. We can use these to kill bacteria implicated in causing disease in humans and animals.

It is important to recognise that there are many different types of antibiotics, and that we must not think of the term 'antibiotic' as a 'catch all'. This is because there are many different bacteria, and different antibiotics target different parts of the structure of a bacteria.

Some antibiotics will simply not be able to kill or prevent the growth of certain bacteria because they have a mode of action that cannot work in that instance. Even so-called 'broad-spectrum' antibiotics cannot kill everything. Antibiotics can also fail to kill bacteria when bacteria have developed resistance to it. This is a completely normal and natural process, however we have caused this to speed up by using antibiotics, selecting for those bacteria that are resistant to them. When this happens in pigs, resistant bacteria can multiply, which could lead to bacterial disease that we cannot treat in pigs. New classes of antibiotics won't be licensed for animal use and will be reserved for humans, so when they no longer work we will have nothing left to treat bacterial infections in pigs. In the worst case scenario, these resistant bacteria could also affect people.

With this in mind, you can help reduce the risk of the development of antibiotic resistance in your pigs by taking the following precautions:

- **Prevent not cure:** Preventing disease means that you will need fewer antibiotics to treat it.

- **Use only when necessary:** Use antibiotics only under veterinary supervision in diseased animals that require it.
- **After reaching a diagnosis:** Reaching a diagnosis means that antibiotics can be as targeted as possible, and the correct one is more likely to be selected.
- **As your vet prescribes:** Using antibiotics at the correct dose, for the correct number of days, and via the correct route minimises the chance of antibiotic resistance developing.

Antiparasitic Medicines

Antiparasitic medicines kill parasites. Different classes will kill different types of parasite and are available in different formulations. Their use is discussed in Chapter 12, 'Parasite Control'.

Hormones

These can be used for fertility and breeding interventions. When a naturally occurring hormone is insufficient (due to disease or other factors) a synthetic replacement can be administered to mimic its natural effect. An example is the use of oxytocin to assist with uterine contraction in a sow struggling to give birth. The use of several hormones is discussed in Chapter 13, 'Breeding'.

Alternative Medicines in Pigs

There is no recognised definition for alternative medicines in the veterinary field, however the term generally refers to interventions where there is not a recognised body of evidence for their use. This does not mean that they do not work, but just that we don't have evidence that they definitely do work at the moment.

The important thing to remember with alternative medicines is that there is less evidence that they will work than conventional medicines, and therefore must not be used instead. Where there is a medicine that is proven to work and is suitable for a particular situation, it must be used. Then as long as it is safe to do so and is prescribed by a vet, alternative medicines may complement this. Before using alternative medicines, be sure to check their legality in food-producing animals.

It is also worth noting that the use of physiotherapy to treat conditions is allowed, but a vet must have seen the pig first to diagnose the condition, and the therapist must then work under the vet's direction.[2]

GIVING MEDICINES TO PIGS

It is fairly common for keepers to apply medicines to pigs themselves, either for preventative health care such as vaccines or worming, or for treatment. Your vet may give an initial dose with you and suggest that you continue the course. Only do this if you feel happy and confident to do so. This guide is not intended to replace veterinary advice but to complement it and remind you of what your vet has probably recommended.

Oral Medicines

Oral medicines for smallholding pigs are split into those that should be dissolved in water, and those that are 'top dressed' (poured) on to food. Make sure that you understand from your vet how a medicine should be given. Oral medicines that are mixed into feed by a feed mill are rarely appropriate for smallholdings. Most oral medicines have been designed with large pig farms in mind, so you will need very small quantities for one pig or a small number of pigs. It is therefore crucial to weigh out the daily stipulated dose with some accurate digital scales.

For medication intended to be dissolved in water, be sure to understand how this should be dissolved, and in what volume of water. Commercial farms will use a dosing pump system that prevents the medication from permanently sitting in the bottom of a tank, but these are unlikely to be feasible on smallholdings. Your vet may be happy for you to dissolve the medication in a small amount of water that can then be mixed with some mashed banana (not from a human kitchen) and given separately to each pig that requires it, but this will depend on the medication and the reasons for its use. 'Top dressing' of medicines on to feed must only be used where products are specifically authorised for this purpose.[3]

Oral medication can only work if the pig ingests it. Sick pigs tend to go off food first, and water second. Therefore monitor your pigs closely so that the intended medication gets into the correct pig, and contact your vet with any changes.

Injections

Giving a pig an injection is the best way to ensure that the intended medication gets into the right pig; however, the procedure also comes with its own complexities.

Handling and Restraint

Before giving an injection, re-read Chapter 6, 'Handling', and ensure that you are setting yourself up for success. The handling and restraint required will depend on the specific pig. A well-handled pig may require just a feed bowl in front of it, others will need to be on a snare. The level of restraint will also vary as to the type and volume of medication; thus it is likely that large volumes of injection or potentially painful substances will require pigs to be snared. With piglets, it is most appropriate to pick them up and another person to inject them with a disposable needle and syringe.

Drawing Up

Always use a clean needle to draw up medication to avoid issues such as injection-site abscesses. Ensure that there are no air bubbles in the syringe, and don't mix different medicines in the same syringe unless they have been licensed for use as a combination product.

Dose

Your vet will inform you as to the dose to use, which will be based on the pig's weight. This is tricky on holdings without weigh scales!

Here Sue is using food to distract this pig for an injection, as well as an injection aid to ease the process.

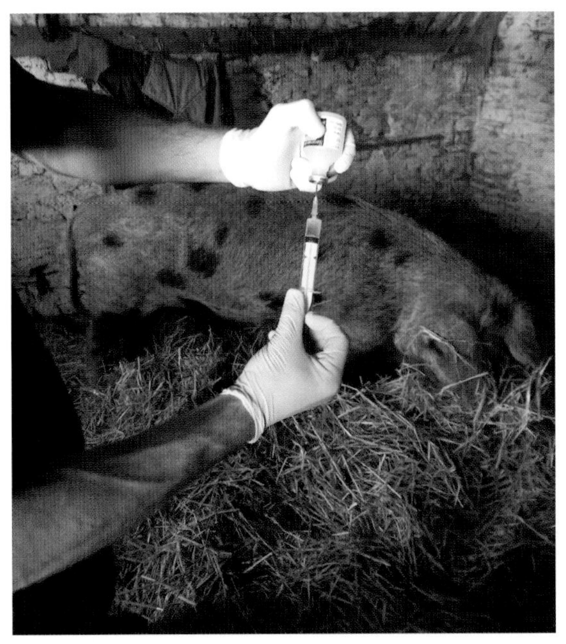

Drawing up medication with a new needle and syringe.

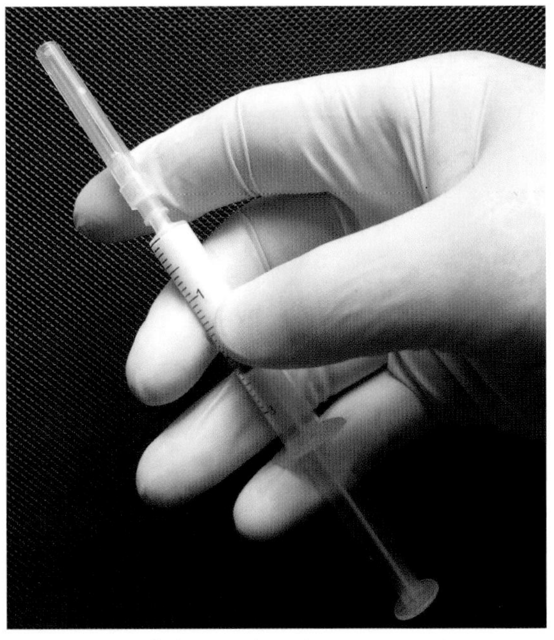

Draw up medicine so that there are no air bubbles in the syringe.

A weigh tape can be useful, or you can follow a formula.[4]

You can also find this formula in a calculator form online from Phin Hall at: https://www.phinhall.net/online-pig-weight-calculator/

Useful Devices

It is very common to break syringes or needle hubs in pigs due to their thick skin and fast movement, and therefore injection aids can be helpful. Devices must be cleaned and disinfected between every use, for example by taking it apart and soaking it in Milton

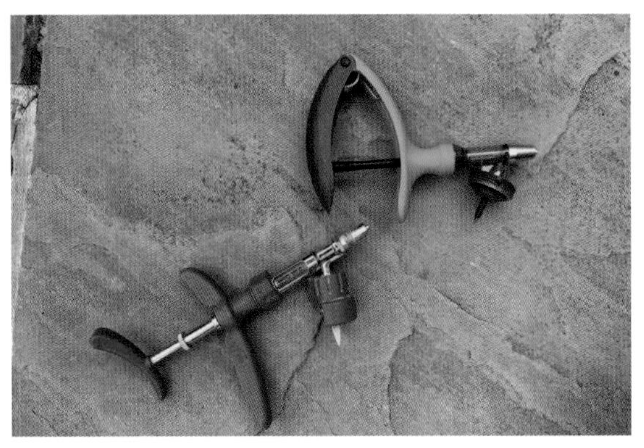

Injector guns can be a helpful aid for injecting several pigs with the same injection, such as when vaccinating.

This formula[4] can be useful to estimate the weight of a pig using body measurements when you don't have weigh scales.

$$\frac{\text{Heart girth (")} \times \text{heart girth (")} \times \text{length (")}}{400} = \text{Weight of pigs in lbs (1lb = 0.45kg)}$$

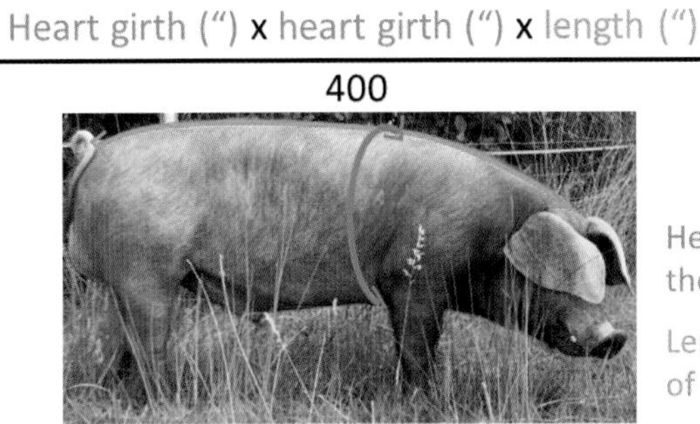

Heart girth = measurement around the body, just behind the front legs.

Length = measurement from the base of the tail to the base of the ears.

sterilising fluid, and brushing it clean. The three most commonly used are injector guns, reusable syringes and slap shots.

Injector guns: Injector guns allow a bottle to be directly screwed on, meaning that the dose will be drawn up after each injection. This is really useful if you are injecting multiple pigs with the same medication, for example six-monthly vaccinations. Putting bottles on to a gun destroys the seal, so only use these when you won't use the bottle again.

Guns tend to be either 2ml or they can be adjustable up to about 6ml per dose. Before use, calibrate the gun by drawing up the dose into a syringe to check that it is correct. After you have pulled the needle out of the pig, hold the gun level and depress the trigger. The capsule can then fill with the next dose - but always check that there are no air bubbles in the capsule and that the full dose has been drawn up before the next injection.

Reusable syringes:[5] Reusable syringes are very sturdy syringes, which prevent breakages and allow you to inject the pig more quickly. They come in sizes up to 30ml. Use the minimum size that you need for the injection, as the larger ones are harder to push down.

Slap shots:[6] Slap shots are plastic tubes that fit between a syringe and a needle to extend your reach. This is so that when the pig moves, nothing breaks in the pig and you can still inject. The side of the slap shot that attaches to the needle has a plastic disc that allows the needle to be 'slapped' into the pig. This does not need to be forceful, it just describes the motion. Inject into the opposite side of the pig from the side you are standing. Once the needle is in, you can then press the syringe and inject hands free, which is less intimidating for the pig. It is really important to note that the tube holds 2ml of liquid, so you will need to draw up extra medication to compensate for this, or follow it with saline. Your vet will be able to advise you on what they would prefer.

Needles

Pigs have a thick layer of subcutaneous fat, which means that needle selection (correct gauge and length) is important to ensure that the antibiotic is given by the correct route. Too shallow an injection will result in low blood levels, effectively under-dosing the pig. Needles should be changed as frequently as possible, preferably every pig. Metal-hubbed needles are useful in pigs, for the same reasons as using injection aids.

The following table can be used as a guide for correct needle size.

Reusable syringes come in different sizes.

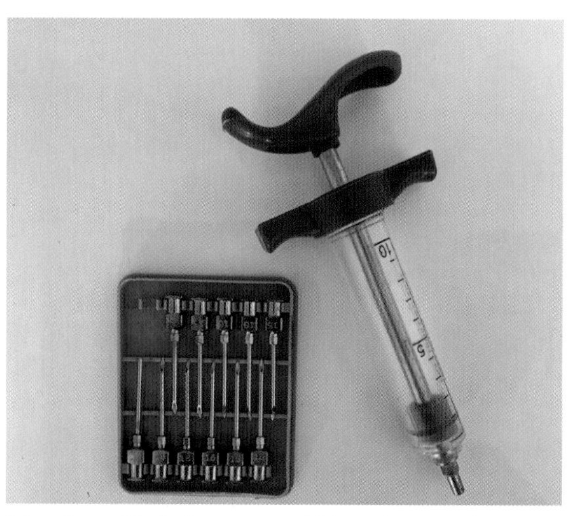

Metal needle hubs can avoid breakages when injecting pigs.

Guide for Correct Needle Size

Subcutaneous injection				***Intramuscular injection***	
Weight (kg)	*Stage of production*	*Length of needle*	*Gauge of needle*	*Length of needle*	*Gauge of needle*
< 7	Pre-weaned	5/8in	21g	5/8in	21g
7-25	Weaner	5/8in	21g	1in	19g
25-60	Grower	1/2in	19g	1in	19g
60-100	Finisher	1/2in	19g	1in	16g
>100	Adult	1in	18g	1.5in	16g

Injection Sites

Pigs can be injected into the muscle (intramuscular), just under the skin into a layer of fat (subcutaneous), or into a vein (intravenous). The appropriate route will be displayed on the medicine. The difference between intramuscular and subcutaneous injections in pigs is that the angle and length of the needle alters. Intravenous injections are only appropriate for your vet to complete.

Pigs should be injected on the neck in the centre of the triangle created by the spinal border, the back of the jaw and the front of the shoulder blade.

For subcutaneous injections, use a shorter needle and angle the needle at 45 degrees to the skin.

For intramuscular injections, use a longer needle and inject with the needle perpendicular to the skin.

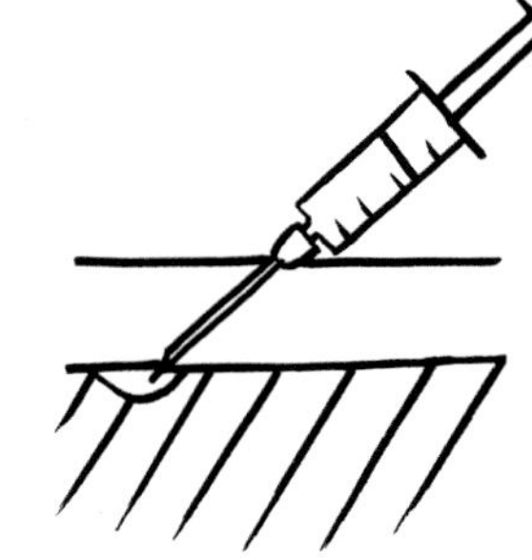

For subcutaneous injections, use a shorter needle and inject at a 45-degree angle.

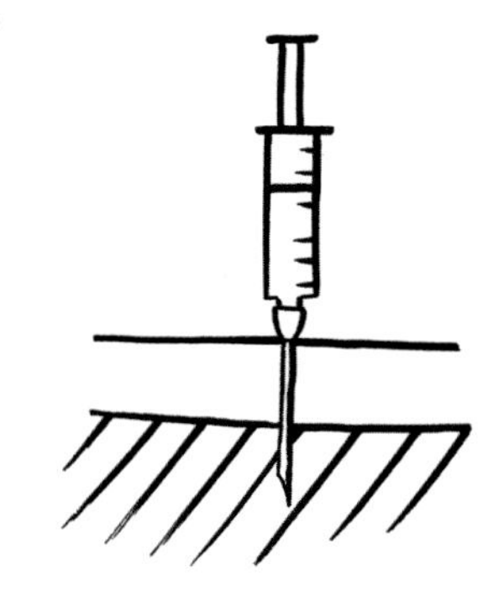

For intramuscular injections, the needle should be perpendicular to the skin.

For intramuscular or subcutaneous injections, pigs should be injected in the centre of the triangle shown, using the anatomical points to mark out the edges of the triangle.

Check the medication guidelines for the appropriate volume to inject into one site, and split your dose accordingly.

To inject, stand next to the shoulder with your pig restrained as required. Feel the area that you want to inject. You will note that it is very fleshy and won't require 'force' in the way you might have seen in cattle. Give the area a clean, push in the needle at the required angle, and inject. Take the needle out with the plunger still depressed (important if you are using a vaccine gun or similar).

A few drops of blood, or sometimes slightly more, is nothing to worry about. The video 'injecting Kunekune pigs' by The British Kunekune Pig Society (copyright Sam Jones) goes through this perfectly: https://www.youtube.com/watch?v=0FeZqzCgX_s&t=9s

In piglets it can be helpful to pull the skin slightly to the side before injecting - using a much smaller needle of course!

Correct injection technique.

This pig is injected without any hassle using the back of a trailer as a form of restraint.

INJECTING INTO THE RUMP

Injecting into the rump of a pig is practised, however it is less preferable than using the neck, for the following reasons:

* Abscesses can occasionally occur at an injection site, especially if reusing needles. In the neck this is inconvenient. In a back leg this could cause a really serious lameness and be more uncomfortable for the pig.
* The sciatic nerve runs down the back leg of the pig. In most pigs it would be difficult to hit, however anatomical differences do happen, and could result in irreversible paralysis.[7]
* This cut of meat is a more valuable cut so is not an area that we want to damage, if possible.
* The skin is thick in the rump, and standing behind a pig generally makes them walk (or run) away from you. Many people find neck injections easier, as you can move with the pig.

Mark the Pig

As soon as you have injected, mark the pig with a temporary identification mark, for example a spray mark, so that you know which pig you have injected. An identification mark (as written in your medicines record) should remain visible on the pig until the end of the withdrawal period, so it may need reapplying later.

It must be very clear from the medicines book which pigs have been injected. Therefore the spray colour should match with what the record says, and there should be no other pigs in that pen with the same spray mark that haven't had the same injection. It is therefore wise to have a good selection of colours! If you are giving several doses, mark on an extra line for each dose. This is so that you can see how many doses of medication a pig has had from just looking at it, and will help you to assess how pigs are recovering as you move around the holding.

Pig crayons often provide longer-lasting marks than spraycans.

Broken Needles

In the (relatively rare) case of a broken needle that remains in the pig after injection, the Red Tractor example broken needle policy[8] should be followed. Immediately mark the pig and spray mark a circle around the broken needle. With the pig on a snare, first use pliers to try to get the needle out. Do not hurt the pig by doing this, you are simply seeing if you can grab the needle. You may prefer your vet to do this instead. If you can't remove it, the pig must be individually marked and details of the event recorded in the medicines record (date, animal ID, injection site).

It is absolutely imperative that this cut of meat does not enter the food chain. Broken needle pigs to be retained for breeding must never be sold on, unless direct to slaughter, and must be regularly health checked. If a pig is to go to slaughter, wait until the withdrawal period for the medication is over. Contact the abattoir in advance to ensure that they are happy with this pig being sent. Spray mark the site of the broken needle on the pig, and include details of the event in the Food Chain Information form.

Accidental Self-Injection

Accidental self-injection is best prevented using methodical and systematic handling and injecting. If it does occur, you must seek medical advice. If you are told to go to the hospital Accident and Emergency department, take the Summary Product Characteristics for the medicine and get someone else to drive you. Oil-based products, such as many vaccines, are particularly dangerous to self-inject and can result in the loss of a finger. A worst case scenario is that a non-sterile needle can lead to sepsis.

CHAPTER 11

VACCINATIONS

Vaccines are specifically used to prevent diseases and are therefore used differently from the medicines that we use to treat them. In fact, vaccines are a route towards decreasing the unnecessary use of other medicines by preventing the disease in the first place.

Vaccines work by exposing the pig to a part of a pathogen (harmful organism) that causes an immune response. After this response 'memory cells' will remain present for some time circulating in the blood, meaning that if the pig meets the pathogen again in the near future, it will be able to respond much more quickly and effectively than the first time, and is therefore more likely to fight off the disease.

The 'part' of the pathogen that we include in a vaccine has been chosen to promote the best immune response possible in the pig, whilst not actually infecting the animal. It can range from a killed version of a pathogen, a live but weakened form of it, or even individual proteins that the pig's immune system will recognise.

Not every vaccine will completely prevent disease, but instead it may reduce the severity of signs experienced, or reduce shedding, meaning that an infected pigs infects fewer other pigs.

WHEN TO VACCINATE

Vaccines can be used in a couple of different ways, which will affect the timing of their administration. Vaccines can be given to the pig that we would like to protect from disease in later life. Generally, an immune response takes about two weeks, and often that response is best after a second dose of vaccine a few weeks later (known as a booster). Over time the effect of the vaccine will decrease; this is why we repeat erysipelas vaccines every six months. It is important that the vaccine course has finished and the immune system has had time to respond before the pig is exposed to the disease. For example, in the case of the reproductive disease parvovirus, we want pigs to be covered for the entirety of their pregnancy

Despite this piglet's hardy exterior, she has an immature immune system so needs protecting with mum's antibodies for the first few weeks of life.

period. Therefore it is important to give parvovirus vaccines at least two weeks before conception.

Alternatively, vaccines can be given to a sow to protect her piglets in their first weeks of life. Piglets are born with an immature immune system, meaning that they don't start to make antibodies until three to four weeks of age.[1] This makes them highly vulnerable to disease in early life. To protect the piglet during this time, a special type of antibody is transferred to piglets through the sow's colostrum, which is the sow's first milk. This means two things: first, vaccine given too early in life is unlikely to work as, instead of causing an immune response, it will be mopped up by the antibodies that are transferred from the sow to the piglet. This is why there is a minimum age at which most vaccines should be administered to the pig, and why it is crucial to discuss these points with your vet.

Second, it means that vaccines given to the sow at the right time during gestation may allow antibodies to be transferred to the piglets to protect them when they drink her colostrum. A really important example are *E. coli* vaccines, which are administered to sows at six and three weeks pre-farrowing to protect piglets from pre-weaning scours.[2] However, for this to work it is crucial that all piglets receive enough colostrum of good enough quality, and quickly enough to protect them.

If the pig's immune system is already busy fighting another disease it will not be able to respond to a vaccine adequately, so only vaccinate healthy pigs wherever possible.

STORAGE OF VACCINES

Vaccines should be kept refrigerated at 2-8°C from purchase to administration. This information can be checked on the Summary Product Characteristics. They must not be frozen, so if taking them home with an ice pack, make sure that it isn't directly touching the vaccine vial. Also prevent freezing by ensuring that vaccines aren't stored pressing against the back of a fridge. If the cardboard packaging is wet when you remove it from the fridge, this is a sign that it has been frozen at some point and could therefore be useless.

Once opened, pig vaccine bottles need to be used in a short space of time because they do not contain preservative. Internet 'hacks' suggesting to put cling film over bottles or other 'solutions' unfortunately do not work and simply put the animal at risk of the disease that you are striving to protect them against. We understand that it is annoying to buy so many bottles, especially as one bottle often treats a larger number of pigs than the number that you are vaccinating. However, per dose, these vaccines are incredibly cheap when compared to the clinical cost of these diseases, even when using a small number of doses.

COMMON PIG VACCINES

The following section will discuss what we consider to be core vaccines, and some common choices of add-on vaccines. By core vaccines, we mean the ones that we consider imperative. All other vaccines can then be evaluated on a case-by-case basis alongside your vet. The following list is by no means exhaustive, and there are hundreds of vaccines available that could be suitable for your specific holding. We will only discuss vaccines that are given by injection, as orally delivered vaccines are unlikely to be practical for smallholders.

Core Vaccines

Vaccination against erysipelas

We have discussed the importance of erysipelas in terms of its commonality amongst smallholding pigs,[3] its severe clinical signs, and its potential to (rarely) cause

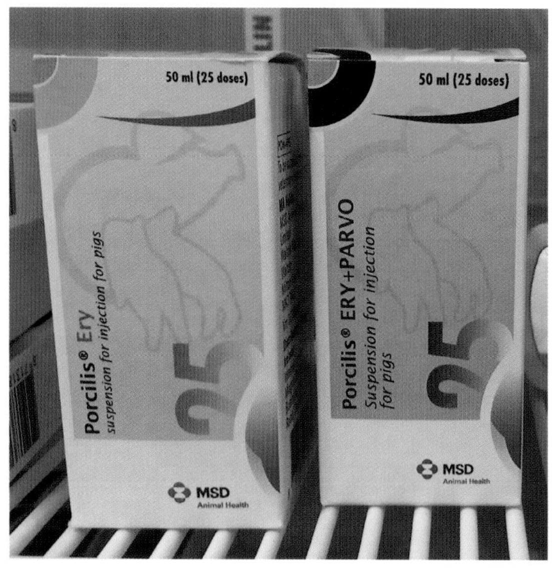

We regard erysipelas as a core vaccination for pigs.

human disease. Common vaccine choices are ERYSENG® (HIPRA) and Porcilis® Ery (MSD), which can be applied from six months or ten weeks of age respectively. Generally, the sow's antibodies will protect piglets up to this point. An initial course entails two vaccines, four weeks apart. Then pigs should be re-vaccinated for erysipelas every six months.

VACCINATING GROWER PIGS FOR ERYSIPELAS

Whether or not producers should vaccinate grower pigs that are just going for meat will depend on your specific holding and how much disease you see. Most producers won't, but speak to your vet if you are concerned that you might have seen cases.

Vaccination Against parvovirus for Breeding Pigs

Due to the commonness of this virus and the fact that it can't be treated, we consider parvovirus a core vaccine for breeding pigs. Usefully, parvovirus vaccines exist in dual formulations with erysipelas, so this doesn't represent any extra jabbing for our breeding pigs. Popular examples are ERYSENG® PARVO (HIPRA) and Porcilis® Ery+Parvo (MSD). These vaccines have slightly different schedules, so speak to your vet about the product that they stock and the schedule that should be used. Ensure that this is completed in good time before mating so that the pregnancy will be protected.

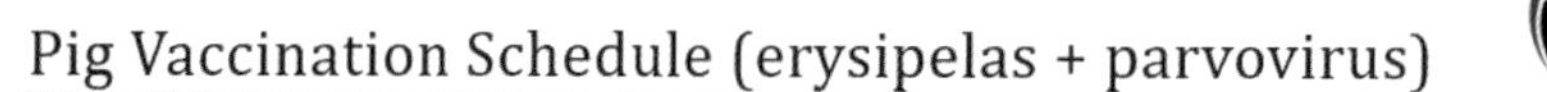

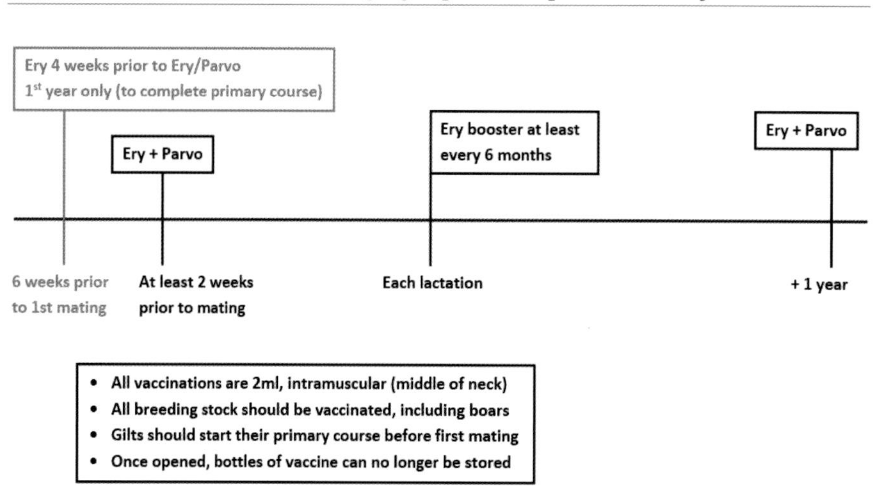

A suggested core pig vaccination schedule.

Once the initial course has been completed, parvovirus vaccines just need yearly boosters. Therefore for breeding pigs you can dose with an erysipelas vaccine once a year, and then six months later (but also once a year) you can use a combined erysipelas-parvovirus vaccine. In this way you can ensure that both diseases are covered. Boars can pass on the infection to sows, meaning that breeding boars should also be vaccinated.

Extras to Consider

Vaccination Against Clostridial Species

Different types of clostridial bacterial disease will cause different signs and affect different ages of pig. To avoid *Clostridium perfringens,* which causes sudden death in piglets, sow vaccines can be used. Examples include Lambivac® (MSD), Entericolix (Boehringer Ingelheim), Gletvax 6 (Zoetis) and SUISENG® (HIPRA), some of which prevent both *E. coli* and clostridial piglet diseases. Speak to your vet if you have concerns that you may have cases on the holding, and be sure to have piglet deaths post-mortemed.

Clostridium novyi[4] causes liver disease and sudden death in sows, and is therefore best prevented through vaccination. It tends to occur around farrowing, and usefully, SUISENG® (HIPRA) also protects against this bacteria.

Vaccination Against Enzootic Pneumonia (EP)

This disease is very common and there is no reason to think it is not present on many smallholdings. Many smallholders do vaccinate for this disease, especially those showing pigs. Stellamune (Elanco) is a popular choice amongst smallholders, but there are many options that your vet will be able to advise you on. EP vaccines only reduce clinical signs, so operating good biosecurity to prevent the disease from entering is far preferable.

Vaccination Against Porcine Reproductive and Respiratory Syndrome (PRRS)

PRRS vaccination is incredibly valuable on holdings infected with PRRS virus, but should not be started without diagnostic testing, proving that the disease is present on the holding. This is because this live vaccine can cause a herd that has not seen PRRS before to experience clinical signs from the vaccine strain. This is also the reason why it is crucial to only use one type of PRRS vaccine on a holding, and pigs vaccinated with one vaccine brand should not meet pigs vaccinated with another.

Furthermore, there are many possible regimes in terms of PRRS vaccines that will be suitable for different scenarios. It is therefore important to speak to a vet who is knowledgeable about PRRS before commencing PRRS vaccines.

Vaccination Against PCV2

PCV2 virus is very common, and vaccination is a mainstay on commercial holdings. This disease is poorly documented on smallholdings, but vaccination is the best prevention if signs are seen.

Vaccination Against E. coli

E. coli causes both pre-weaning diarrhoea and post-weaning problems of meningitis, diarrhoea and sudden death. Vaccines for pre-weaning diarrhoea are best delivered by vaccinating the sow. Many commonly used clostridia vaccines also confer protection for *E. coli.*

Vaccination Against Many Infectious Diseases

Many infectious diseases can be vaccinated against, from leptospirosis (now licensed in the UK in a dual combination with erysipelas and parvovirus by MSD) to ileitis. Further options can be discussed with your vet.

CHAPTER 12

PARASITE CONTROL

To understand parasite control, it is crucial to understand the significance of the environment in terms of harbouring parasitic eggs, as well as understanding that parasites go through multiple life stages. In order to remove a parasite you must tackle both the environment and the pig, as well as every life stage of the parasite. If one link in the chain of these life stages is not broken, the parasite will persist. For these reasons, many holdings merely control parasites rather than prevent or remove them.

PARASITES DEFINED

Internal parasites include worms and single-celled organisms such as coccidia. External parasites include mites and lice. Ticks and fleas rarely cause problems in pigs.

Parasite control in pigs is not nearly as sophisticated as for sheep and cattle, where there are detailed plans to prevent the development of resistance to our commonly used anti-parasitic medicines, and far more options to choose from in terms of treatments. We have much less need for multiple therapeutic options because, amazingly, worms in pigs do not seem to show the same resistance. There is currently just one documented case in the UK of pig-worm resistance, which was Oesophagostomum resistance to the ivermectin class of antiparasitics.[1] This case has raised the need for potentially more judicious use of wormer, especially ivermectins.

First we will discuss control of internal worms and external parasites, and then control of coccidiosis.

HYGIENE AND LAND MANAGEMENT

Good hygiene and land management can be considered the mainstay of internal parasite control, with therapeutics used to augment this. It should focus on reducing pig contact with their own faeces, and therefore parasite eggs. Sue and Stephen bring their pigs inside for worming, collect the faeces and put them on a special dung patch that is not spread. This avoids wormer (and resistant worms) contaminating pasture.

Using the same pasture continuously for pigs for many years may allow a resistant population of worms to build up. Commercial units will use land for a maximum of two to three years, rotating with crops. Organic producers will do this roughly every six months to maintain grazing and prevent worm burdens from building. Even if you bring your pigs indoors over winter, we know that some worms can survive a mild winter[1] so pigs may be reinfected when they go back out.

Management of Outdoor Small-Scale Pig Housing

Management should involve implementing the following measures:

- Move pigs on to new ground frequently, and rest and resow pasture between.
- 'Poo pick' where moving on to new ground is not possible.
- Use rotational grazing with other species if possible.

Management of Indoor Small-Scale Pig Housing

Indoor systems can be problematic, as faecal oral contamination of indoor pig housing is likely to be very high and parasitic eggs can be very difficult to kill. Management should involve implementing the following measures:

- Fully clean out areas, removing poo exactly as you would for a dog.
- Ensure that housing is totally dry after cleaning out.
- Try to achieve all-in, all-out systems if possible.

PREVENTING THE SPREAD OF EXTERNAL PARASITES

External parasites are very tricky to control once on a holding. In order to eradicate external parasites from a holding, every pig must be treated at the same time with an avermectin medicine (ivermectin or doramectin), and this should be repeated until no pigs have any parasites still living on them (as medication cannot kill egg stages). This is sometimes not feasible due to the withdrawal time of these medicines, meaning that some pigs cannot be treated, and the practicalities around injecting all pigs. There is also likely to be some environmental contamination, making this procedure even more difficult.

The pig louse, Haematopinus suis.

Therefore you are far better to ensure that no pig with an external parasite ever makes its way on to your holding by considering this during the isolation period. If you have some pigs with external parasites remember that scabies can't live for long off the pig, so leave housing, arcs and paddocks empty for at least a week between pigs to avoid spread between pigs on the holding.

THERAPEUTICS

For pigs, there are two classes of antiparasitic medicines available: the benzimidazoles and the avermectins. As always, use of these products must be under direction from your vet, who can decide best courses of action based on your specific holding.

'WORMERS'

The word 'wormer' is often used synonymously to 'antiparasitic medicine'. However, it is important to note that some of our 'wormers' also deal with ectoparasites. When we are just referring to worms, or helminths, the most appropriate word that we are looking for is 'anthelmintic'. However, 'wormer' is still the most commonly used term in the field.

Benzimidazoles

Fenbendazole and flubendazole can be found in products such as Pigfen® (Huvepharma) and Flubenol™ 5% (Elanco), which are oral products that can be mixed with the pigs' food. They are capable of killing worm adults, larvae and eggs so can be used as a singular dose for most worms. They do not treat ectoparasites so are useless for itchy pigs.

Avermectins

The avermectin group includes ivermectin and doramectin, which are found in products such as IVOMEC® (Boehringer Ingelheim) and Dectomax™ (Elanco). Avermectins also kill ectoparasites, so their use is indicated in the itchy pig or to prevent isolated pigs bringing lice or mange on to a holding. They do not kill egg stages and therefore repeated doses are required two to three weeks apart. Generally the oral forms are not suitable for smallholdings as they are only available as premixes.[2] Neither treats liver fluke, but this is very rarely found in pigs. They are also not licensed to kill *Trichuris suis* (whipworm).

We have found a very low amount of ivermectin resistance in Oesophagostomum,[1] and therefore it is advised to preferentially use a benzimidazole when an avermectin is not specifically indicated (that is, when ectoparasites are not present or intended to be targeted).

When to Use Antiparasitic Medicines

Use of antiparasitic medicines can be informed by frequent faecal egg counting (FEC), at least every six months, or can be performed routinely. FECs do not detect external parasites, which are best prevented from coming on to a holding with good biosecurity. Once they are present on a holding, a protocol must be decided alongside your vet based on the clinical signs that are seen.

Faecal Egg Counting (FEC) for Internal Parasites

A FEC involves submitting a sample of fresh faeces to your vets, an external company, or even doing this yourself if you have the tools. The sample is mashed up and added to a solution, which causes the parasite eggs to float. The solution is then put into a special microscope slide that allows the number of eggs per gram to be counted. The decision as to whether to treat with an antiparasitic medicine is based on this number.

A FEC is very useful to inform treatment decisions; however, there are some problems with its use in pigs:

- *Ascaris suum* or 'milk spot' is ingested and migrates through the pig's lungs and liver, causing disease. After about twenty-one days the mature worm reaches the gut and lays eggs. It is only

Smallholder Jamie collecting a faecal sample for FEC.

Samples for submission should be labelled with the farm name, the date and the test required.

at this point that eggs can be detected in faeces with FEC. Therefore it is very possible that a new infection from *Ascaris suum* could cause very severe disease, but an animal would have no eggs detected at FEC.[2]

- *Trichuris suis* is only capable of laying eggs (that could be detected by FEC) several weeks after infection, and adults only intermittently lay eggs. It is advisable to worm grower to adult pigs showing signs of diarrhoea. It is also important to note that the type of diarrhoea caused by *Trichuris suis* (often bloody) is a red flag for the much feared disease swine dysentery. Therefore alongside worming, it is important to contact your vet.
- The standard method of FEC doesn't pick up heavier metastrongylus eggs. These are quite rare, however, and would cause coughing.

The specific number of eggs per gram that will warrant treating with an antiparasitic will be different depending on the holding, taking into account concerns such as pasture contamination. The specific pigs that should be wormed (all versus a particular group) will also vary. Therefore these factors and a course of action should be considered alongside your vet.

COUGHING PIGS

We suggest worming pigs that start coughing. In addition you should worm if you receive reports of 'milk-spot liver' from the abattoir. For pet pig owners, a post-mortem in the case of any deaths would allow the liver to be assessed for any 'milk spots' and could give clues about other pigs on the holding.

A Sensible Strategy

In addition to routine FECs carried out at least six monthly, or routine use of antiparasitics where decided alongside your vet, there are some additional times that dosing with an antiparasitic medicine may be advisable:

Pigs entering the holding: These might include a new pig, a boar back from loan, or animals back from show. During the two-week isolation period you should dose twice (fourteen days apart) with an antiparasitic that will kill internal and external parasites (an avermectin). If a pig is seen to be itchy during this period, speak to your vet about whether further doses and a longer isolation period may be necessary.

Pigs experiencing clinical signs of parasites: These might include the following:

- Itching, ear crusts, head shaking
- Coughing
- Diarrhoea in growers to adults
- Weight loss or poor growth rates
- Observation of lice and/or anaemic piglets

Pigs leaving the holding: To ensure that you do not introduce parasites on to another holding, dose twice (fourteen days apart) with an antiparasitic that will kill internal and external parasites (an avermectin).

Sows infected with lice or mites before farrowing: Dose with an antiparasitic medicine to lower the risk of her passing these to her piglets.

Responsible Antiparasitic Use

Just as with antibiotics, we must use antiparasitic medicines appropriately in order to reduce the risk of resistance:

- Ensure the dose is calculated for an accurate weight, using the prescribed product and the correct route of administration.
- Use antiparasitic products only when necessary.
- Avoid bringing parasites on to your holding by sourcing pigs from trusted sources and enforcing a strict quarantine.

CONTROLLING COCCIDIOSIS

Most commercial pig holdings administer toltrazuril-based products to prevent this disease at three days of age. Historically this has been given orally, but more recently this has been formulated into injectable products also containing iron to be given at the same time. On most smallholdings this will be unnecessary due to the length of time between litters, because fresh paddocks can be used, and equipment can be thoroughly cleaned and rested; instead, environmental control can be practised. Control could take the form of the following, but should be decided with your vet who has an understanding of your specific holding:

- Use toltrazuril-based products for clinical cases, remembering the long withdrawal of these medicines.
- Where outside, move farrowing arcs on to fresh ground for each litter, and clean and disinfect arcs and equipment with an oocidal (coccidial egg stage killing) disinfectant.
- Where inside, clean and disinfect housing and equipment with an oocidal disinfectant, ensuring that everything is totally dry afterwards.
- Control rodents, which are significant spreaders of coccidial eggs.

CHAPTER 13

BREEDING

Keeping pigs all year round and dealing with the intricacies of pregnancy and farrowing is a big decision. Similarly to sourcing weaners, find a breeder who will mentor you through the process (this might be different to the one that was suitable for weaners). Good records and an organised system will be key. Be sure to have kept pigs over a winter period, and be confident that your land will cope with a much larger burden.

Breed your first litter in the spring, remembering that your sow will need to be mated four months earlier. This is the time that she will be most fertile, and also means that you are less likely to be managing a lactation during very cold or very hot weather.

Before breeding pet pigs, you must consider whether you have excellent homes for all the piglets. Rescue and sanctuary organisations are full of pet pigs where homes could not be found. You must also consider castration, and confirm that you have a vet in the area who is happy to do this in the way that meets your requirements.

SOURCING BREEDING STOCK

If you do not intend to keep your own boar (which can be challenging for new breeders) you need to consider boar hire. Buying gilts locally may preclude you from using the seller's boar as he will be the father of their gilts, so you can either buy local gilts and source a boar from afar, or buy gilts from afar and use a local boar. That being said, an experienced pig breeder, especially if they are interested in pedigree stock, will know their own bloodlines and may be able to provide gilts and boars that are guaranteed to be unrelated to each other.

Decisions around purchasing breeding stock should be informed by the reasons that you want to breed, as described below.

To produce meat weaners: In this case your gilt/ sow will need a good back end to take the weight of the boar, but her conformation and underline don't need to be perfect, as long as she has enough teats to rear piglets and she is up to breed standard. It is less important if the boar and sow are more closely related than ideal, but it still comes with risks of poor health.

To raise breeding pigs: Boar and sow should be at least up to breed standard, but entirely unrelated. The breed society/ BPA will do a free kinship analysis for you.

To show: Your pigs will need to be perfect. They must have an absolutely even underline with, ideally, fourteen teats. Study the breed standard very carefully before buying, and take advice from an experienced pig shower if you can.

To help the breed: In this case, contact the breed society and ask what lines need help and if there are any that need spreading across the country. The society might even be able to help you with travelling costs.

To raise as pets: Temperament and reducing inbreeding is absolutely key here in order to breed pigs that can have happy and healthy lives.

REGISTERING PIGS

The offspring of two registered pedigree pigs are not automatically eligible to be registered themselves. Whatever the intended fate of the piglets, the litter should be birth notified with the BPA, but only those that fit the breed standard can then be registered. The registration can only be carried out by the breeder, so once a pig is sold it is too late. In order to ensure that there are only great examples of each breed producing more pigs, it is important that you don't breed from anything that isn't registered.

What to Buy

Your final, biggest decision is whether you buy weaners, gilts or in-pig sows to start your breeding enterprise. The cheapest way to get pedigree breeding pigs is as weaners. These will probably be seven to eight weeks old, and will have been checked for basic conformation and underline, but there is no guarantee that they will grow into the prize-winning pigs you might have hoped for. The other disadvantage is that you will have to feed them for the best part of a year before they can be put in pig.

Buying 'gilts' means buying young female pigs that are almost ready to breed. This will give you a better idea of their adult conformation, but they will be more expensive than weaners. Both the above options lead to you taking a chance of the pig being fertile (infertility is unusual but does happen) and being a good mother. Gilts sometimes don't understand what is happening when they farrow and can reject, lie on, or even eat their piglets.

In-pig sows are the most expensive but most reliable choice. You will be dealing with a sow of known adult conformation and with experience of farrowing, and you will also be able to have piglets much faster, and delay sourcing a boar. The downside is that you will have less time to prepare for farrowing the sow and weaning and selling the piglets, and you will be farrowing a sow that might not know you well, so could be unpredictable.

SELECTING FOR BREEDING

Later down the line, once your herd is established, you will need to decide whether to breed your own replacement gilts, or buy in new stock. This will depend on the genetic diversity on your holding, how long you have been using the same boar, or whether you use artificial insemination. Close genetic matings are likely to cause birth defects and are also not allowed to be registered as pedigree. This means that good records will need to be kept to ensure that close matings are not set up accidentally.

In order to maintain piglet numbers in your sows, you should not breed your sows over seven litters. Also be aware that fertility will decrease in a sow that hasn't been regularly bred. When too old, as in all species, sows will stop breeding completely.

Home-Bred Gilts

When you choose to select gilts for breeding will differ on different holdings. However, it is important to have half an eye on this right from the day that piglets are born. Pigs from litters with birth defects such as hernias, cleft palates or congenital tremors are not suitable for breeding.

A cleft palate such as this may not affect the piglet's ability to drink, so it is likely that it would make it to selling weight, but none in the litter should be bred from it.

This pig has a hernia so is unsuitable to breed from.

Gilts should receive a good amount of human contact so that they are easy to handle. They should not be kept alone, so it is best to keep on three (in case one dies) at a time for breeding so that they can stay together for life. You don't want them to be fat, so feed sow food when you remove them from the rest of the group. This should be completed by the time they are 40kg at the latest (or less for a smaller breed) to ensure that you will not get brother-sister matings.

You will need to consult your breed society to understand what to look for in your gilts, but in all gilts you should look for the following:

- Conformation: the gilt should demonstrate even weight bearing, easy movement with no stiffness, with the toes pointing forwards and even claw growth.
- Teats: ideally there should be fourteen productive teats. This involves looking properly at each one, so teach your gilts to lie down with a belly rub. Teats can show all kinds of problems, including blind or inverted teats, so these need to be properly examined. Registered pigs should have teats that are in line with each other in pairs.
- Temperament: a bad-tempered sow will make your life difficult in cases such as a difficult birth and when you need to

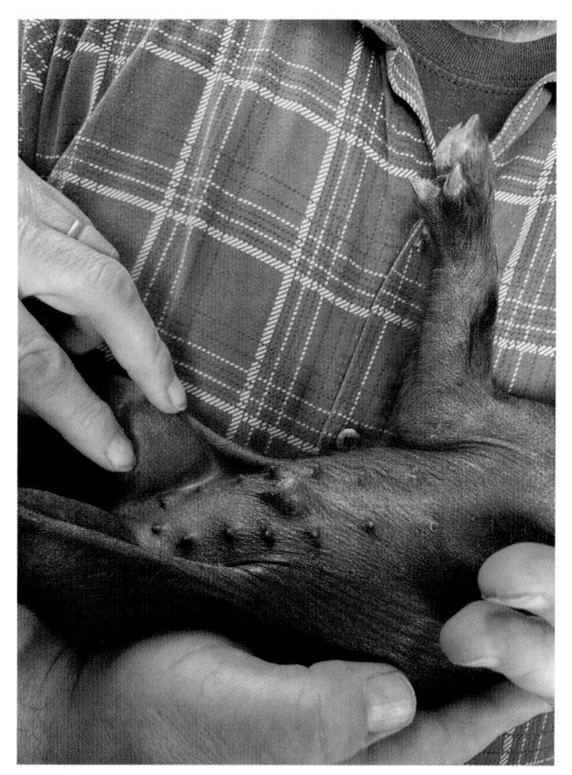
Checking a four-day-old piglet for sex, underline, and number of teats.

assist her, or for doing anything with her piglets later on. This does need to be balanced with mothering ability, as fiercer sows can be more protective of their piglets and can be better mothers.

Bought-in Gilts

Make sure that you understand the vaccination status of incoming pigs and match it with your own during the isolation period. If you are not absolutely sure of a pig's consistent vaccination schedule, you should restart the course. Consult Chapter 3, 'Biosecurity for Pig Holdings' for more information on managing this isolation period.

Acclimatisation involves a slow introduction to the diseases on your farm so that gilts can overcome any infections before they are required to cope with the stress of a pregnancy. Remember that new pigs coming on to the holding are not just a risk to your pigs: your holding's diseases may actually be a risk to new pigs! This time also allows vaccines given during the isolation period to take effect, which is especially important if you have had to restart courses.

The easiest way to achieve acclimatisation is to put your new gilts into a paddock alongside other pigs to allow nose-to-nose contact. Historically, placenta from another sow would be put in the pen, but this is now illegal and must not be done.

HEAT CYCLE AND SERVING

Gilts should be a minimum of 220 days before service, and should not be mated on their first heat cycle on smallholdings. They are generally about ten to twelve months when they are first mated, remembering that you won't be able to register piglets born to a sow younger than one year.

The heat cycle comes round about every twenty-one days, but this can differ between sows and seasons by a few days. Sows are fertile for about three days, peak fertility occurring about twenty-four to thirty-six hours into her heat.

After a sow has had piglets, her next cycle will be determined by when they are weaned. Around five days after weaning a sow will normally come on heat. This can be less

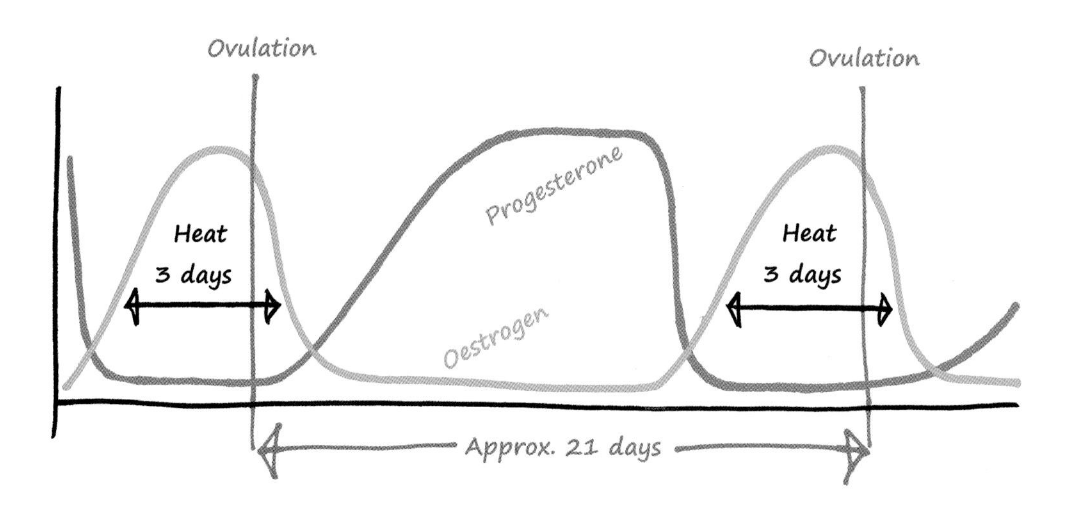

The timings of a sow's heat cycle.

reliable on smallholdings, as later weaning can mean that sows can come on heat before weaning, especially if a boar is nearby. The process can also be disrupted by a stressful event (such as moving), meaning that a sow comes on heat unexpectedly early. This means that smallholders may need to be better at detecting heat than commercial farmers.

Signs of Heat

Vulval swelling (puffing up) and reddening are the main signs of heat. The sow will also appear more boisterous and noisy, initially riding other sows and being interested in boars, and later standing to be ridden. She may have a clear discharge at her vulva, and may flick her tail.

'Standing heat' will occur when she is ready to be mated. This can be tested by pressing

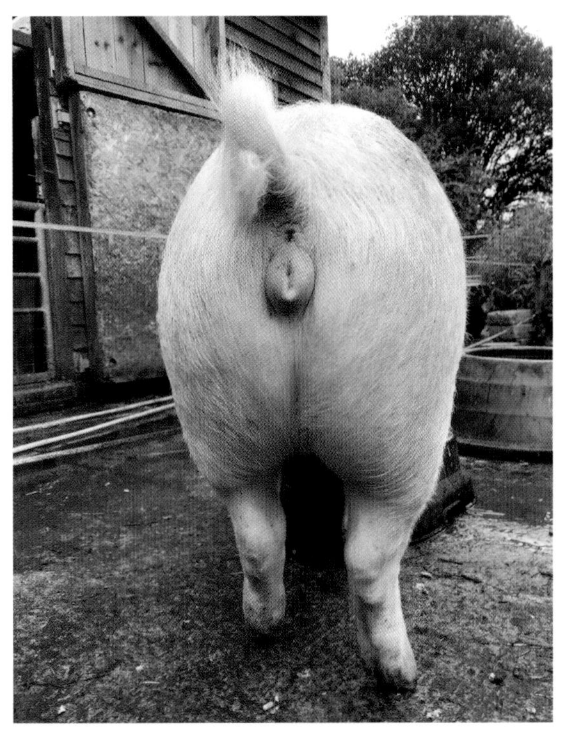

Some very clear signs of heat in a gilt.

The vulval changes of a heat cycle can even be easily detected on some black pigs like this one, but certainly not all.

Pete is testing this sow for heat by applying gentle pressure to her back.

down on her back over her rump. If she is in 'standing heat' she will stand rigidly. She is most fertile twenty-four hours into standing heat.

Whenever you notice a sow on heat, write it on your calendar. Knowing when your sows will next come on heat will greatly assist your organisation of the holding.

SERVICE

Before serving, ensure that the sow is ready to become pregnant in terms of body condition score. You may wish to use a technique called 'flushing', which involves increasing her feed intake for fourteen days before service to increase litter sizes.[1] Ensure that all vaccinations and parasite control are in place several weeks before service.

Smallholders generally choose natural service over artificial insemination (AI), which involves putting a tube of semen into the sow yourself. The suitability of AI versus natural service will depend on your specific situation.

Keeping your own boar may be unsuitable if you have very few sows as he is likely to be left lonely for extended periods, which, as well as being a poor welfare state, may leave him frustrated and difficult to manage. This can be avoided by keeping a castrated male or an old and infertile sow with him for company. It will also only be a matter of time until he will be set to mate his children, so boar rotation is critical. In addition some environments, such as care farms, will not be able to house a boar.

Most smallholders will opt for natural service.

Boars can be hired, however this has the potential to bring unwanted diseases on to your holding even when practising good isolation techniques. Be sure to source a boar with a good temperament, and consult Chapter 6, 'Handling', for specific advice. Also be mindful of who you lend your boar to: not everyone takes the same level of care.

Instead of hiring a boar, sows can be sent away, which can be a good option for those fairly inexperienced with boar handling and breeding. Remember that their return constitutes a disease risk, and you won't be able to closely monitor them while they are gone. You will want to know when the sow has been mated, so send them to an experienced breeder. They will also need to be carefully reintroduced to pigs on the holding to prevent fighting.

AI centres thoroughly test boars to ensure that diseases will not be transmitted through semen, meaning that disease risk is much lower. However, smallholding AI can have lower success rates. Furthermore, smallholders looking for a specific genetic line of boar may struggle. For most breeds there are a couple of boars available and they are rotated frequently.

Commercial units will bring gilts into heat at the same time for their first insemination with a synthetic hormone. Following this first litter they will wean at four weeks and inseminate when sows come on heat five days later.

Match boar and sow size so that you don't get oversized piglets, and to avoid sow injury in the case of natural service.

Natural Service

Boars should be at least ten months old before being used for service. Boars and sows need to be introduced very carefully. Boar tusks can be very dangerous and can cause serious injury to the sow. Fighting and injury will be much less likely if the sow is content with a boar's advances, so introduce them when the sow is on heat.

Sows are territorial, so introduce the sow to the boar rather than the other way round. Sue and Stephen take the sow out of her pen, put the boar in, then let her back in after a couple of hours. This also allows the boar to spend a few hours marking his territory and will make the area smelly and appealing to the sow. Ensure that introductions take place in a large area so that they can move away from each other if required.

Watch what happens closely, but don't interfere. Pigs take around fifteen minutes to mate. A boar's penis is corkscrew shaped and twists as it engorges with blood. This process must reverse before he exits, or serious injury can occur. The process requires a large volume of semen to be ejaculated, up to 500ml, and finally the boar produces a plug of jelly that block's the sow's vagina and prevents semen from leaking out. Therefore it is imperative that he is allowed to complete the task!

Observation of service is absolutely ideal as this allows you to plan.

If you see him mounting the sow and she doesn't moving away, write down this date so that you can track your dates and prepare for the big day. Boars are generally hired out for six weeks, so if the sow doesn't 'take' first time he gets another chance.

If you don't see any action or she doesn't become pregnant with a boar that you know is fertile, speak to your vet. Sows can become acyclic but can sometimes be brought back into heat with a hormone injection. Only rarely are sows completely infertile.

Artificial Insemination (AI)

Deerpark will be the best place to go to for rare breed semen, but there are many other breeding companies offering more commercial AI. The BPA and Deerpark provide an excellent 'how to' guide.[2]

Ensure that you have a really good grasp of your sow's cycle so that you will order semen at the right time. Timing will be your best aid to successful insemination. Sperm stays alive for twenty-four hours in the sow so you don't want to inseminate too early. Nor do you want to inseminate too late and miss ovulation.

Order semen as soon as you see signs of heat, or when the sow will come on heat if you have been able to get an idea of her heat cycle. Serve twelve hours after she first comes on to a standing heat, and then again twelve hours after that. You want to aim for two inseminations before she ovulates. If she is not still in standing heat by the second time you go to serve, the next time around serve straightaway and then again in twelve hours, as you may have missed ovulation.

You will generally receive non-frozen semen for AI, which should be kept as close to 17°C as possible. Ten minutes or so before inseminating, bring it up to body temperature by putting the bag close to your own body heat.

A safe area for AI is crucial for both sow and stockperson. A small, outdoor, clean pen is fine. The floor should be non-slip, which can be assisted by laying down some straw. Most smallholders will be doing AI because they don't have a boar so won't be able to use the smell of a boar to excite the sow and make her more likely to suck in the semen. If you are using a boar, you must ensure that they are separated by a fixed metal pen so that he can't inseminate her.

If you don't have a boar, or if you have a lazy boar, you can fill a pen with boar pheromone spray. Always move the gilt to the boar or the boar pheromone pen rather than the other way round. If you are working outdoors, spray a towel with pheromone spray that she can sniff. At this point, check for standing heat again.

Hygiene is absolutely crucial to avoid introducing infection when performing AI. Use clean lubricant and only put lube round the rim of the catheter rather than on the end. It can be spermicidal so shouldn't come into contact with the semen (a pre-lubricated catheter is best as it will be sterile). Use a dry paper cloth to wipe her vulva and clean off any dirt, being careful not to push it into her.

For the next part of the process, gentle handling and patience is critical, which is aided by polite and well-handled sows. Insert the catheter slightly to the left and pointing up by 45° until you feel the cervix stop your slow advance. Then, twist the catheter slightly to lock it into her cervix in an anticlockwise (left) direction.

Attach the semen to the catheter and hold it above the sow. Do not squeeze the bag or bottle: she needs to draw in the fluid naturally over the next few minutes with no rush. If semen does not flow in at a reasonable rate or is coming back out of the vulva, stop and start again right from the process of ensuring that she is definitely on heat. When you are happy that the semen is flowing in, rub her flanks and push your knee into her side to simulate what a boar would be doing. This will improve oxytocin release, will cause her to suck in the fluid more efficiently, and will improve the chances of conception. Finally, leave her in the same space for at least fifteen minutes. Whenever you remove a catheter remember to turn it back in a clockwise direction to undo the lock that you have created.

DIAGNOSING PREGNANCY

Knowing the date of insemination will assist management of the pregnant pig and preparation for farrowing. If you saw the covering you can have a fairly accurate idea of when your pig is due to farrow, all going well with her pregnancy. Gilts especially may not show signs of pregnancy until the final month or later, so this may be your best opportunity to prepare.

Pig pregnancies last for 115 days, which is also three months, three weeks and three days.

Monitor the sow very closely either twenty-one days after you saw them mate, or twenty-one days after you inseminated her for the signs of heat. If you didn't see them mate, monitor twenty-one days from when she was on heat. If she comes on heat, she is not pregnant. If she does not show signs of heat, she may be pregnant. Be sure to keep a close eye throughout the pregnancy. If she loses the pregnancy she may come back on heat at an irregular time.

Scanning

Scanning is the most definitive way to tell whether a sow is pregnant at a particular moment in time. This is achieved using an ultrasound scanner at about twenty-eight days. Much earlier than this won't be detectable. Experienced scanners will be able to detect pregnancy right through to farrowing, but less experienced scanners may struggle much later than forty days. Some vets are happy to do this, or a professional pig

This pig is being scanned with Doppler ultrasonography to find out if she is pregnant.

Pete's colleague Tom scanning a sow with visual ultrasonography.

scanner can be hired. If you are selling sows as in-pig, we recommend getting scanning photos to show this is the case. This gives you and your buyers assurance that you are selling pigs that were at least pregnant at some point.

A Doppler scanner can be used but is more prone to operator error and won't give you a photo. Furthermore, if the sow is on heat you may get a false positive. If the boar has been left with the sow, he may have served her later that you thought, so it is unwise to place too much reliability on ultrasound findings. Positive means positive, but negative doesn't necessarily mean negative!

Visual Signs of Pregnancy

Early in the pregnancy, some keepers will look for a change in the angle of the sow's clitoral hood for pregnancy diagnosis, but this is not reliable. Otherwise, visual signs of pregnancy include:

- A larger-looking belly from eighty to ninety days. The sow can eventually look full and the abdomen slightly dropped. This can occur earlier, from about sixty days, especially in older sows.
- Mammary development about two weeks before farrowing: the sow will develop an

This pig is only eight weeks pregnant, but a tell-tale line is starting to show on her flank.

Beatrice is seventy-four days into her pregnancy here and you can see an 'udder line' just developing. This is the faint line between her flank and her udder. Her udder is also starting to develop.

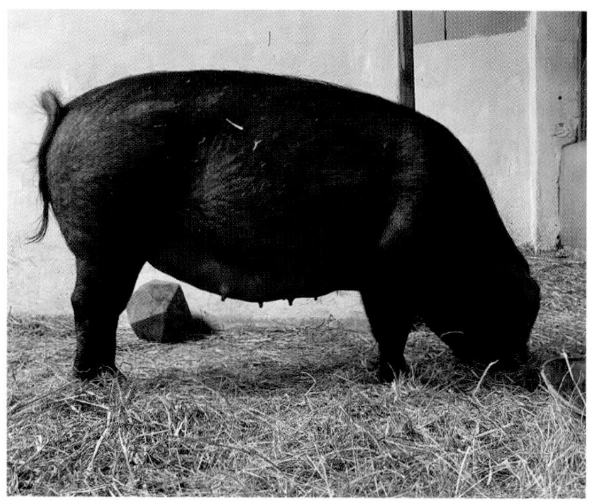

This second-litter sow is twelve days from farrowing. She is more difficult to tell but her tummy looks quite full and there is some mammary development.

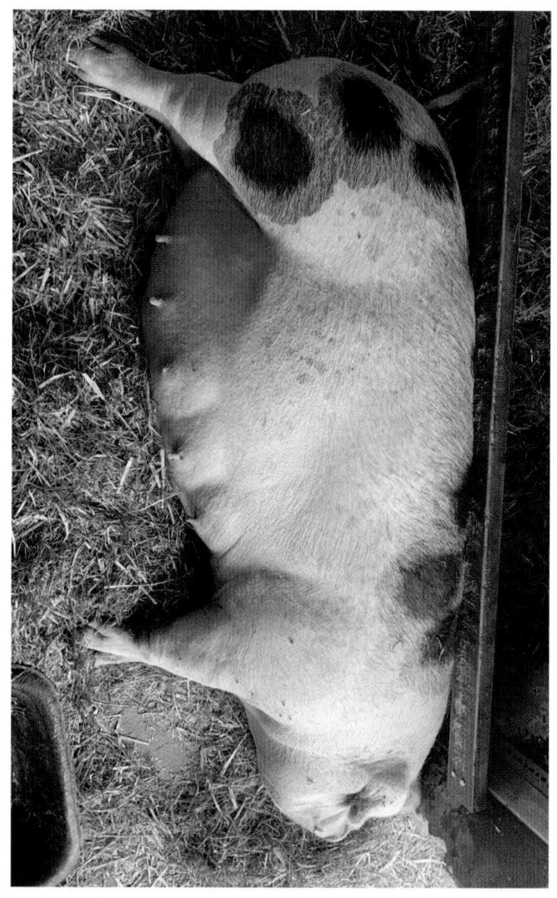

At 116 days into her gestation (she was slightly late) you can now see that each mammary gland has developed and her teats are filled.

This gilt has a developed udder and her teats look full. She farrowed later that day.

extra section of tummy with teats and udders.

- Vulval swelling and redness (in pink pigs) around four days before farrowing.
- Teats will fill with colostrum twenty-four to forty-eight hours before farrowing. Around six hours before, colostrum may flow easily if a teat is gently squeezed, or might just drip from the teats. If it is just

Annoyingly, pigs don't always read the text book. This sow was producing milk eight days before farrowing.

dripping, be sure to collect this to feed to your last born and smallest piglets who may miss out on this valuable colostrum!
- Nesting marks the start of stage one labour.

MANAGING THE PREGNANCY

It is important to minimise stress throughout a sow's pregnancy. Don't move her or mix her with other pigs between six and thirty-five days into the pregnancy, as she can easily reabsorb the pregnancy. Preferably also don't give injections between these times.

Be aware that any illness causing a high temperature can lead to abortion. Sows will generally abort if they have fewer than four piglets, although this is less reliable in rare breeds, which tend to have smaller litters. Ensure that feeding is correct, with regular body condition scoring, as described in Chapter 5, 'Feeding'.

Be sure to give any vaccines necessary to support the piglets in the six weeks or so before farrowing, checking the manufacturers' guidelines. It is also sensible to worm any sows who are infected with external parasites, as discussed in Chapter 12, 'Parasite Control'. Ensure that your farrowing accommodation is fit for purpose, re-read Chapter 4, 'Pig Housing and Environment'.

Move the sow into her farrowing accommodation one week before she is due to farrow at a body condition score of 3 to 3.5.

Pregnancy Problems

Problems with a pregnancy can range from early loss of the embryo to abortion (expulsion of the pregnancy at less than 109 days into gestation). Each problem has its own set of possible causes, so the more you know about your pigs and the history of the pregnancy, the more your vet will be able to narrow down the list of options.

If a sow comes on heat twenty-one days after mating, either she did not get pregnant, or embryos were lost during the very early stages. If she comes on heat thirty-two days after mating, she was pregnant but four or fewer embryos were present at around day seventeen, as the sow will generally not support fewer than four embryos at implantation stage, so they were absorbed. In both of these cases there was a problem during early pregnancy. Conversely, a mummified piglet means that the embryo has died after thirty-five days, which suggests a problem later on in pregnancy.

That being said, investigating reproductive problems can be really frustrating in pigs. Pigs can lose a pregnancy during any stress, and the majority of abortions are not due to infectious disease. It would be foolish not to consider infectious disease as a possibility and allow disease to continue to move through a herd, however many reproductive

The abortion of an eight-week-old litter, for which there are many possible causes.

disease investigations serve to rule out the possibility of infectious disease, rather than diagnose a specific cause.

In terms of investigation for infectious disease, aborted or stillbirth material can be submitted for laboratory analysis. Put it straight on ice and call your vet to inquire whether this is advised for you. Mummified piglets tend to be less useful. Your vet may also want to run other diagnostic tests, such as some blood tests.

REPORT POSSIBLE NOTIFIABLE DISEASE

Notifiable disease must be considered in cases of severe reproductive disease, so always contact your vet if you have a sick sow with abortion or dead piglets, or if multiple sows are affected.

Be sure to review your vaccination schedule and ensure that everyone (including boars) has received their erysipelas and parvovirus vaccines, as well as any others that you and your vet have implemented. Next, assess the body condition score (BCS) of sows, checking that all sows are farrowing at BCS 3–3.5 and that none are above 3.5. Then review general management, such as their housing and environment, or handling. Check that you are not causing stress in sows from six to thirty-five days after insemination, which could prevent implantation.

Review your insemination process, especially if you are using AI. If you are using natural service, is your boar tried and tested? Is he as active as he once was? Has he been sick? Sperm production is a process that takes six weeks and requires a cool temperature (which is why testicles are held outside the abdominal body cavity), meaning that fertility problems are common after a boar has been sick.

Sows experience seasonal dips in fertility. Historically the sow would breed just one litter per year, in the spring. Sows are therefore most fertile at Christmas and least fertile in early summer, with a slight dip also in autumn. We have tried to reduce this to be able to breed pigs all year round, but it is still apparent. Generally you will see an increase in sows not getting pregnant or aborting their litters over the summer. This should be managed by keeping sows as cool as possible during hot weather in the summer months.

Clover and mycotoxin are further possible reasons for reduced fertility, which your vet will be able to discuss.

FARROWING

An indoor or outdoor farrowing area must be constructed as discussed in Chapter 4, 'Pig Housing and Environment'. It must be absolutely clean, and faeces must be removed religiously. A dirty farrowing area will lead to diseases later on such as joint ill, septicaemia, diarrhoea and meningitis. The other imperative is plentiful water.

Farrowing begins when the litter starts to become stressed due to piglets being restricted inside the uterus. They release the

This farrowing area is ready to house the sow.

stress hormone cortisol, and this leads to another hormone to be released by the sow, called prostaglandin F2a. Prostaglandin F2a release leads to oxytocin release, which is part of the hormonal pathway that leads to the uterus contracting and the piglets being expelled. The birth process occurs about twenty-four hours later. A small litter will lead to less piglet stress, which is why smaller litters often farrow late.

A sow farrows in three stages. An awareness of these stages allows us to judge whether a sow needs assistance.

Stage One Labour

The sow will start to make a nest eight to twenty-four hours before farrowing. Anything available might be incorporated, so ensure there is nothing around that might be dangerous. That being said, it is really important to provide plenty of nesting material. Straw is most sensible, but clothes from washing lines may also be at risk! Let her do what she needs to do, and remove what needs to be removed once farrowing has started.

Nesting can be quite dramatic! This sow has removed all the straw from the floor to leave bare concrete and there is a feed bowl buried in it.

From this point onwards, the sow can become much more aggressive. Don't take chances, and keep a pig board with you at all times.

This is the same bed with excess straw removed so that piglets don't get trapped in it later. Sue and Stephen cleared even more straw once farrowing had started.

Straw hasn't been removed from this nest, and therefore piglets are quite likely to become caught up in it and squashed.

The sow will be restless, and her udder will be full and producing colostrum. Her body temperature will have risen. She should eat less, but not stop eating, and should drink more. Sows that stop eating may be at increased risk of the disease post-partum dysgalactia.

Stage Two Labour

Stage two labour is characterised by piglet birth. During pregnancy piglets are held in the uterus, across two huge horns. At birth, they are born randomly from these horns. The whole process should take two to four hours in total.

OXYTOCIN RELEASE

Contraction of the uterus leads to pressure from the unborn piglets. This causes oxytocin release, which leads to stronger uterine contractions, more pressure, and therefore even more oxytocin. Therefore, this oxytocin release is critical for the farrowing process. It is also critical for milk let-down and expulsion of the placenta later on.

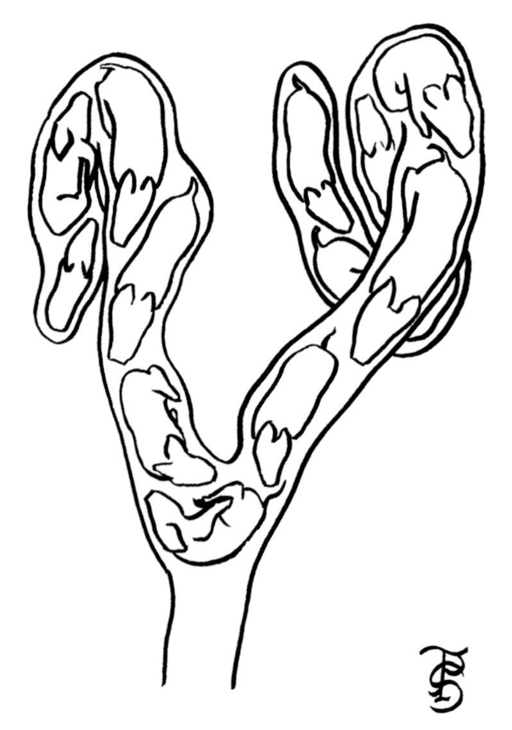

Piglets are delivered in random order from each horn, and will be delivered either head first or breech.

The vast majority of farrowings take place without any need for intervention. Piglets are far smaller than the offspring of many species in terms of their size relative to their mother, and therefore pushing them out is not as much of a struggle.

Sows will generally lie on their side, and piglets will be born either head first or breech (backwards). Breech presentation is rarely a problem in pigs as the piglet is so small compared to the sow so rarely becomes stuck. Straining is present but is less dramatic than some species. The sow's tail will twitch as she delivers each piglet, and a piglet will be born every fifteen to twenty minutes, sometimes

Piglets will be born every fifteen to twenty minutes.

This first piglet has literally just been born, and has already found her way on to a teat while she waits for her brothers and sisters to join her. The oxytocin stimulated by this immediate suckling is directly assisting with the birth of the other piglets. Nature is wonderful!

This male piglet is being born backwards, but without difficulty.

with a break of up to one hour after the first few piglets.

If healthy, piglets are very mobile and quickly find their way around. Some piglets may need a little more help, so if necessary, ensure that their mouths are free of placenta. If they need prompting to breathe, rub their flanks and put a bit of straw slightly up their nose, exactly as you would a lamb. You can rub them dry, warm them and show them a teat.

Sows will often suckle the piglets between piglet births, which should be encouraged to start colostrum intake, especially over long farrowings. You may observe piglets passing

Gently helping a newborn piglet to suckle.

meconium (brown faeces). This indicates that the piglets have sucked sufficiently, as milk is an excellent laxative and it shows that the digestive system is working.

It is normal for umbilical cords to be intact immediately after piglets are born - they will break naturally, and seal in the vast majority of cases. Occasionally a cord will need to be broken manually, by gently pulling it away from the sow so that the cord is as long as possible from the piglet. Sometimes it may need tying or pegging to stop bleeding, but milk the blood in the cord towards the piglet first; piglets can't afford to lose more than about 5-10ml of blood!

This piglet has been delivered caught up in the first section of placenta, so needed to be freed.

Unfortunately, stillborn piglets such as this one are fairly common.

The placenta from one horn will often be delivered partway through farrowing as one horn is emptied, so do check that there isn't a piglet caught up in it. The birth of dead, stillborn piglets is fairly common (about 7 per cent of piglets), especially the last in each horn of a large litter.

Most keepers will not dip navels in iodine as the umbilical cords are so long that the cord dries and drops off before any infection has time to make its way along the cord. However, if joint ill (piglet lameness and joint swelling) is experienced, then dipping the severed end in 10 per cent iodine solution should be trialled and hygiene should be improved. Also be careful that cords don't become tied up and tangled in straw.

Once farrowing has finished the piglets should suckle then fall asleep on the teats, with mum gently grunting to them until she, too, drops off.

This piglet was struggling to break the umbilical cord so Sue and Stephen helped with a gentle tug to break the cord's attachment.

The placenta, or afterbirth, being passed normally.

The piglet's umbilical cord can be dipped with a 10% iodine solution to prevent joint and navel infection, but most smallholders will not find this necessary.

Stage 3 Labour

The sow will cleanse, meaning that she will eject the placenta (the large sac that has held the piglets) within about four hours of the last piglet. The placenta is also commonly called the afterbirth, or cleansing or foetal membranes. Laying the placenta out to ensure that both horns are intact will confirm that cleansing has been successful. It will be very large. Sows will often eat it, but if they don't, remember that it needs to be disposed of in an approved way.

Some vaginal discharge in the few days after farrowing is normal, but keepers should watch out for signs of the disease metritis.

FARROWING PROBLEMS

It is always appropriate to seek veterinary advice before assisting with farrowing problems. If intervening with a sow or piglets farrowing in an arc, particular care must be taken. This is best performed by your vet. The following list of farrowing problems is not exhaustive but includes the most common. If you are worried, please call your vet.

Not Giving Birth

If a sow has not started to give birth when expected, it is most likely that her dates have been recorded or calculated wrongly, or that she is slightly late. It could also be due to dead piglets, which won't instigate farrowing as they cannot release the necessary hormones. If this is the case, your vet will need to use a hormone to terminate the pregnancy. In order to avoid disasters whilst doing this, firm knowledge of the insemination date is essential.

Stressed Sows

Occasionally a sow will be very stressed during labour, especially a gilt. You should only go into the pen if it is safe to do so. Stressed pigs might well accidentally squash a piglet as they thrash up and down. Here a hooked stick is useful for pulling piglets out of the way and into the creep area. Often, once the piglets start suckling, oxytocin kicks in and the sow will settle down. Just occasionally an inexperienced mother might start eating her piglets, in which case move them away to safety and keep them in the creep area. Light sedation from a vet can be required, so please do call for assistance if this happens.

Interruptions to the Farrowing Process

There are times when the farrowing process does not proceed as expected, and intervention may be necessary in the following scenarios:

- Delays of more than one hour in the first half of the farrowing.
- Delays of longer than half an hour in the second half of the farrowing.
- The sow is no longer pushing but you think that there are more piglets.
- The sow is straining but a piglet is not being produced.
- The total farrowing has been longer than four hours.

One of two situations is likely: the sow is suffering uterine exhaustion, or there is an obstruction.

Uterine exhaustion: Larger litter sizes will make sows more likely to become tired and dehydrated. Generally, the older the sow, the less tone she has in her uterus, and the harder it will be for her to push out the final piglets. Fat or skinny sows are also likely to have problems, and hot weather can have an impact.

Obstruction: Piglets can get stuck, especially if the boar used was not well matched to the sow in terms of size. Or, two piglets, one from each uterine horn, may be attempting to come out at the same time.

Intervention

Intervention should always be conducted under veterinary supervision: this can be directly in the form of a vet visit, or it can be through discussion and training before or during the delivery.

Internal Examination

First the sow should be examined internally to check for obstruction. This should not be completed without discussion with your vet as to whether this is appropriate, as damage to the fragile birth canal can easily be caused.

Most uterine horns are out of reach of the human arm so we may not be able to pull piglets out. However, we can generally reach the point where the horns meet each other, which is where a piglet can commonly get stuck.

Internal examination must be scrupulously clean to avoid introducing infection. It must also be safe, and completed in a well-lit room with a sow that is calm and lying down. If there is any risk to you or her, you must call your vet.

A clean, full-length glove must be worn with plenty of lubricant, preferably by someone with small hands. The hand should be slowly and delicately inserted into the vagina, parting the lips first so that dirt is not taken in. If the glove becomes contaminated, take it off and start again. Gently follow the sow's anatomy, not forcing your way in, pointing your fingers towards the spine and feeling for any obstruction. If you do find a piglet, very gently draw it out by the legs or head along a path of no resistance.

A farrowing sow should only be examined with a long glove and plenty of lubricant.

A piglet that has been stuck is likely to struggle to take its first few breaths and may not make it. Clear any material from its mouth and nose, and gently poke a piece of straw up its nose; rub the piglet to try to stimulate breathing. If this doesn't work, swing it briefly from its back legs and pat its chest.

If you can feel a piglet but can't pull it out, call your vet. After an internal examination, always speak to your vet about what medication is necessary, in terms of pain relief and antibiotics. Be sure to tell your vet if there is blood, or if there has been any particular trauma.

Medical Assistance

If there is no blockage, an oxytocin injection may be indicated. This is a prescription-only medicine so must be administered under veterinary supervision. However, we are aware that keepers occasionally administer this medicine themselves, so we will discuss some points to watch out for when using it.

Oxytocin will cause the uterus to contract and therefore must not be used if an obstruction might be present. Irreparable damage would be caused if the uterus is forced to contract with something blocking it. Therefore, veterinary assistance to confirm its suitability is always wise.

Oxytocin is quite commonly overdosed, which can cause a spasmic uterus and potential rupture. It should be used in a way that simulates its biological release, which is in waves, by giving a small amount (0.25ml) and waiting for twenty minutes. At this point it can be given again if necessary. It can be given up to four times, but you should ask for a vet to visit after a maximum of two doses, no matter your level of experience, as something more may be required.

Birth of Dead Piglets

If you have an increase in dead piglets, or more than you would expect, your vet will want to look at whether or not these were born already dead (signifying a problem with the pregnancy), or if they died whilst they were being born (signifying a problem with the farrowing). Unfortunately you can't tell definitively whether or not a piglet dies before or after birth without performing a post-mortem, but you can get some clues by just looking at the piglets. Piglets are born with little slippers on their trotters, which prevent the piglet from traumatising the sow internally. These quickly rub off as the piglet emerges alive, so their presence points towards a piglet that was stillborn. A piglet that has died before birth is also likely to have a long, wet navel. Black spots found on piglets imply that there was some foetal stress during the farrowing.

If the piglet dies after birth this may point towards a husbandry issue that you can assist with, such as an overly long farrowing because the sow is too fat. Deaths before birth should be investigated as a problem with pregnancy.

The Placenta is not Passed

If the placenta has not been delivered more than about six hours after the last piglet, intervention may be required. However, this can be very difficult to judge as the sow may have eaten it. Only if required, the steps that should be taken are exactly the same as for intervening with farrowings, as a piglet that is stuck could still be preventing the placenta from being passed. Passing of the placenta is also under oxytocin control, so oxytocin can be helpful once an obstruction has been ruled out.

THE IMMEDIATE POST-FARROWING PERIOD

The period immediately post farrowing is crucial. Tiny piglets can become cold, dehydrated, or lack energy. This makes them less likely to move out of the way of the sow and may lead to them getting squashed. The sow is also at risk of an infection, either from her womb (metritis) or her udder (mastitis).

Sow and piglets need to be monitored collectively and treated as one entity. Noisy piglets may tell you that a sow isn't producing enough milk and therefore may be sick. If this situation is allowed to progress those piglets will quickly start to experience severe

This sheep fleece provided an innovative way to keep these piglets warm where a heat lamp wasn't possible due to lack of electricity.

Healthy, well-fed piglets are a joy to behold.

The typical appearance of a piglet that has been lain on; this is especially likely for small or weak piglets.

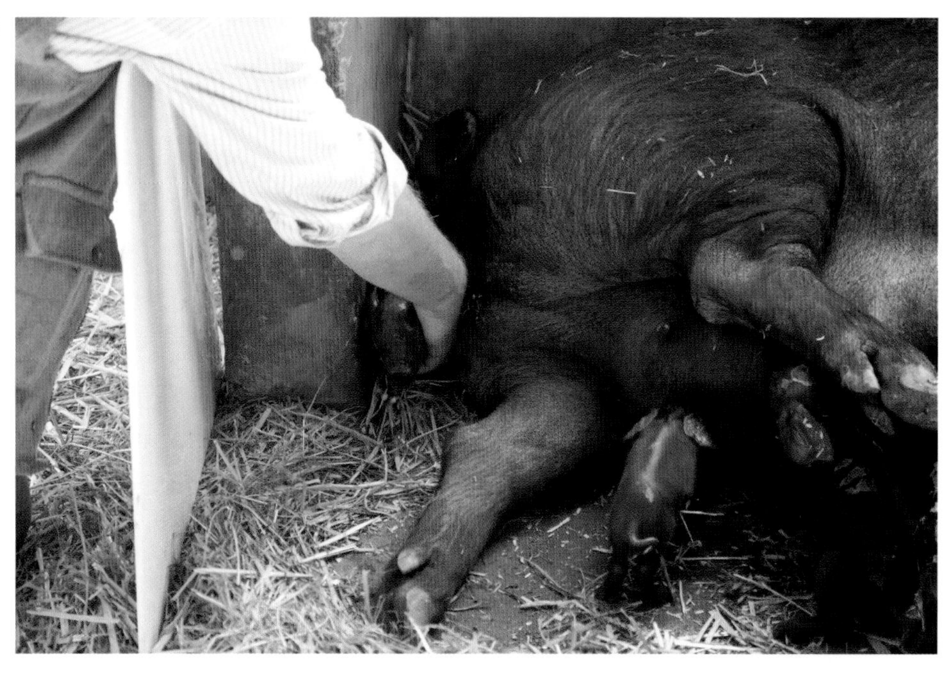
Here a lost piglet is carefully removed from a confined space.

signs of disease. A sow that is acting 'off colour' will stop producing enough milk, quickly leading to hungry piglets. Piglets that are not suckling frequently will also lead to the sow not being able to keep producing milk.

Care of the Sow

The sow must be drinking a large volume, and must be able to do so in a short space of time. She may decrease her food intake for the first day after farrowing – but keep a close eye that her appetite returns by the next day. She certainly should never stop eating as she needs a great deal of energy to make enough milk. Providing tasty, soaked food can help. She should be lying on her side to nurse every hour or so (not on her front so that the piglets can't get to her udder), grunting to encourage the piglets and seeming well in herself.

As long as everything is well, you should leave sow and piglets alone to suckle where possible, so as not to interfere with colostrum intake. To begin with the sow will do little other than nurse the piglets and sleep, and

This sow and piglets require very few interventions.

Stephen taking sensible precautions to check the udder of this sow.

she might well foul nearer to her sleeping area than normal to avoid leaving the piglets. Make sure you keep the farrowing area as clean as possible without unduly disturbing the family.

Even the softest, friendliest sow can completely change character when she has piglets, so take no chances. They normally go back to their friendly selves after about a week when the hormones have settled down.

After this initial period of six to ten hours, check that the sow has enough functional teats for all piglets, but bear in mind that it can be hard to strip milk from a sow unless she is being stimulated by the sucking piglets. A basic clinical examination is advised daily by assessing her demeanour, her food and water intake, and her faecal output. Monitor the udder for changes in skin colour and temperature. Each gland (teat and surrounding area) should feel soft and warm (but not hot), with no swellings. Any pain could be a sign of mastitis (infection of the mammary glands). If something worries you, take a rectal temperature to provide your vet

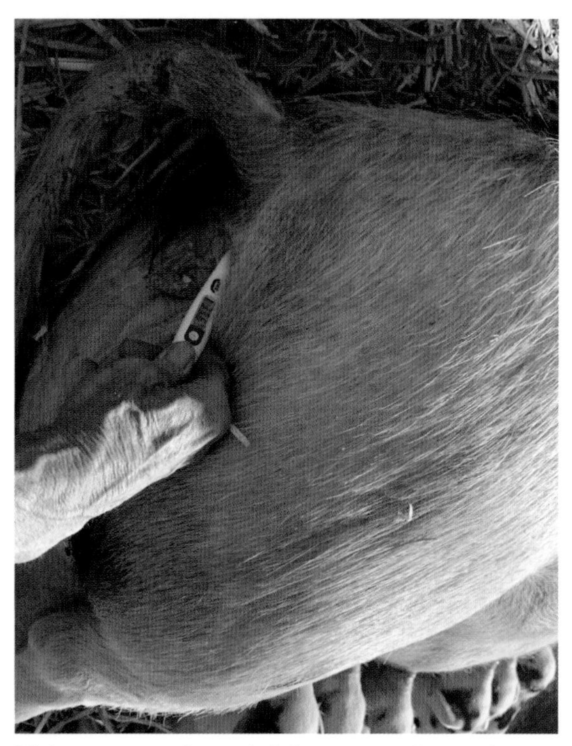

This sow experienced sickness post farrowing with a very high temperature at points. Her keeper noticed this promptly, and with veterinary intervention was able to save her life.

with even more information. When taking the temperature of the sow, note any abnormal smells or discharges that may be emanating from the uterus. A healthy discharge post farrowing should be clear and have no smell.

If you are worried about the sow, it is really important that you call your vet. The conditions mastitis or metritis can be very serious for the sow and need urgent medical attention. Postpartum dysgalactia syndrome (PDS) - or metritis-mastitis-agalactia - whilst not being life threatening for the sow, will quickly lead to sick piglets.

Sow feeding is incredibly important in order to keep up with the huge milk demand over the lactation, and should be gradually increased to as much as she will eat after farrowing.

The Importance of Colostrum

The mantra of 'breast is best' holds true for piglets. The first milk, colostrum, is vital for the piglets' immune system development and is also rich in fat and protein.[3] Piglets are born without energy reserves so will quickly fade away without this nutrition, especially if they are cold. Furthermore piglets don't start to produce their own antibodies (which are important to fight diseases) until three to four weeks of age.[4] In many mammals, antibodies from the mother will travel through the placenta into the unborn animal, however placental impermeability makes this impossible in pigs.[5] Therefore colostrum must provide antibodies until the piglet starts to produce these itself.

COLOSTRUM'S SPECIAL ANTIBODY

Colostrum contains a special type of antibody (IgG) that can be absorbed through the piglet's gut wall into the blood during the first twenty-four hours of life, at which point these channels close.[3]

A wandering piglet can be carefully steered back towards the teat using a smooth stick.

Because of this a piglet must be seen to be drinking its colostrum requirement during the first day of life, which is a minimum of around 165g (about 160ml) per kilogram of bodyweight.[3, 6] This should start thirty minutes after birth. After farrowing is over, nursing will occur approximately every hour.[7] Piglets generally choose a teat and stick to it, so watch out for piglets that are 'teat surfing' in a desperate hunt for a productive teat. When suckling productively their tails will flick and they will emerge from the teat with milk around their mouths.

The piglet 'ritual' of massaging the udder and touching the sow's snout is really important for the sow to release oxytocin and let down her milk. If the litter is not doing this effectively, you may be able to massage the udder to assist this, if it is safe to do so. Following this, the period of productive suckling is actually very short, between thirty-five and sixty seconds.[6] A piglet that

Stephen is checking that every piglet is suckling. This can only be done if the sow is placid and relaxed.

A well-fed piglet will have a plump belly.

Well fed piglets will suckle and then fall asleep.

has suckled will have a full, round and plump belly.

Risk Factors for Low Colostrum Intake

There are several risk factors for low colostrum intake to be aware of, the following in particular:

A large litter: This is less of a problem in rare breeds, which tend to have much smaller litter sizes than commercial sows, but it is possible that some teats are not productive. Larger litters lead to more competition at each productive teat, and therefore to a lower volume of colostrum per piglet; also the production of colostrum decreases after about twenty-four hours.

Small or weak piglets: Smaller piglets can get bullied, leaving them with no or worse teat access. Generally, the first few pairs of teats are the most productive. Also, smaller piglets are normally born later. Piglets born later are also at high risk of insufficient colostrum intake due to the decrease in

The serious nature of this cleft palate would probably prevent this piglet from being able to suckle properly, and so it is likely that it will not receive the colostrum it needs to survive.

Sue and Stephen's outdoor piglet area, useful for piglet interaction and feeding, and which the sow cannot access.

colostrum quality over the first day that it is produced.[3]

Weak piglets simply don't have the energy to compete and suckle productively. Piglets with deformities may need longer to suckle without competition. Finally, if the piglet is too weak to stimulate the teat, the teat will shut down and the piglet will be left without any milk.[6]

Sow diseases: Sow diseases such as PDA, mastitis and metritis must be monitored and acted upon.

Early farrowing or dripping colostrum: Sows that farrow early will produce less colostrum;[3] knowing the sow's insemination dates will assist in the management of this. If a sow is dripping a great deal of colostrum before farrowing, collect this and feed it to the piglets most at risk.

> **SHORTAGE OF COLOSTRUM: WHAT TO DO**
>
> If you are worried that the whole litter or individual piglets are not receiving enough colostrum, you must act quickly using the strategies to combat low colostrum intake detailed below, before the gut channels close at twenty-four hours after farrowing.

Strategies to Combat Low Colostrum Intake

Split Suckling

This technique is suitable if you have a larger number of piglets than the number of functioning teats but the sow is still producing a good amount of colostrum, or if you can see some smaller piglets being bullied off the teat. Mark up the piglets into three groups: the big ones, the mediums and the smalls. The smalls should maintain teat access at every feed, but the big ones and the mediums should alternate access so that one group is contained in the creep area per feed. Alternating between these two groups should allow everyone to be able to drink.

Individual Piglet Care

If you find a cold piglet that isn't moving much but is alive, place it on a warm hot water bottle wrapped in a towel. Dab piglet kick-start product (or honey) on its tongue if it can swallow, or rub it into its gums if not. Providing piglet kick-start may be a sensible precautionary measure for piglets that are small, weak, or have a domed forehead (due to being squashed in the uterus), and can be repeated at eight hours of age alongside some colostrum replacer.[7]

Once the piglet starts to warm up, a drink of warm milk substitute (or colostrum if in the first twenty-four hours of life) will help the process. Pig colostrum replacement is best in terms of immunity and survival, while cattle or sheep colostrum follows that so more should be given to meet requirements.[3] Supplementing with milk rather than colostrum will mean that a piglet does not receive the special antibodies in colostrum, making it more likely to suffer from infectious diseases in the coming weeks.

A bottle is necessary before two days of age. Those for puppies work well, or a 10ml syringe will also work. There is a danger of milk going into the lungs, which can be fatal. To avoid this, only bottle feed piglets with a suck reflex. If they do not have a suck reflex and you have tried kick-start or honey, call your vet as they are very sick. Please do not allow these piglets to fade and die. If they are suitable to bottle feed, hold the bottle at a shallow angle, mimicking that of a sow's teat. Feed about 30ml at a time to avoid them becoming lethargic and risk breathing in the milk instead. This will take them about one minute and forty seconds to drink.[8] If they aren't sucking enthusiastically, stop.

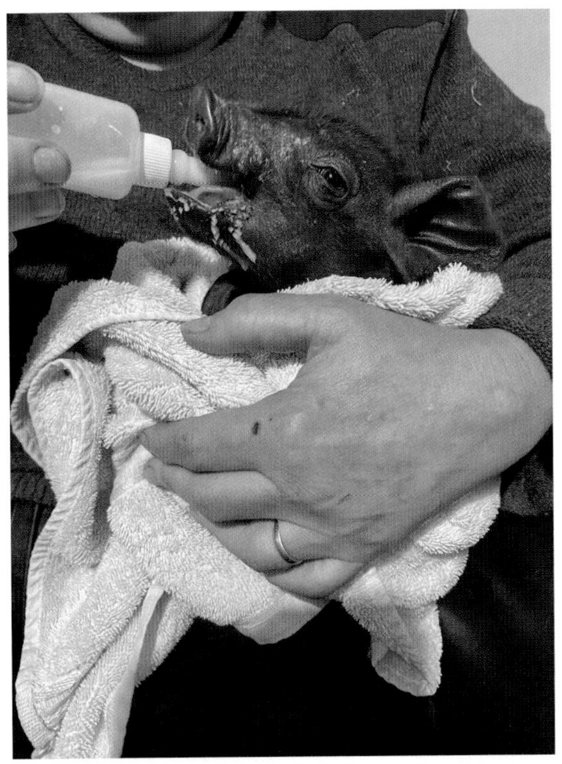

Bottle feeding a piglet; note the shallow angle of the bottle. Sue and Stephen find that swaddling the piglet in a towel so it can't move and then holding it upright seems to be the easiest way to get it to take milk.

Sows tend not to reject piglets, so you can generally carefully reintroduce them even after a couple of days.

Similarly later on, if the piglets are small, weak, or you have worries about the amount of milk your sow is producing, you will need to give them supplementary milk. Fast-growing piglets gain 70g of weight per day,[3] and milk intake must keep up with this. Piglets can lap from about two days old, so give them some milk substitute in a shallow tray whilst mum is eating. If the sow can reach it she will finish any left when she returns. The milk should be given, ideally warmed, at least twice daily. This should be provided for a minimum of ten days.

PIG MILK REPLACER

It is a good idea to buy some powdered pig milk replacer, such as Farramate, before farrowing. However, if you are caught out, goat's milk mixed with natural yoghurt can work well. Cow's milk tends to cause diarrhoea so should be avoided.

A sow will rear her piglets 'her way', and generally will know best.

At some point over the first couple of days, check that the piglets are all able to defecate normally. Atresia ani is when the piglet has a layer of connective tissue over the anus, which will prevent this. You should consult your vet with any condition that causes you concern throughout the lactation, such as a piglet struggling to breathe or not moving freely.

ORPHAN OR HAND-REARED PIGLETS

Hand-rearing orphan piglets is not for the faint-hearted and should only be attempted if a real disaster has taken place, such as the death of a sow at farrowing. Baby piglets are entirely dependent on you, and are therefore a considerable burden. Success will depend on how closely you can mimic the sow's normal mothering behaviours. If you have any doubts about the amount of time you are able to commit, especially during the first two weeks of life, then euthanasia may be the kindest option.

If the sow has died during or shortly after farrowing, colostrum is required immediately. This must be administered to all piglets using a bottle with a small rubber nipple, as previously discussed. This can be sourced directly from a placid suckling sow or as a commercial replacement product. Colostrum from the piglets' own mother will always be best, but this is often impossible to retrieve, and in an emergency even lamb colostrum replacer will be better than nothing.

Once the piglets are fed with colostrum they should be placed in a heated (~30°C) creep area. A traditional kitchen with a range cooker makes a charming alternative to this, but take extreme care if you have other pets in the house.

Gently warm the milk replacer to body temperature and give each piglet as much as they will comfortably drink. For the first three days feeds should be provided as close to every two hours as possible, including through the night, using vocal cues and grunts to make piglets aware it is feeding time and to stimulate them socially. This is

Hand-rearing a piglet is a round-the-clock task.

Despite the tiring nature of handrearing piglets, this one made it to breeding standard!

Unless you have a team of helpers, it will be helpful to set up a small nipple feeder as shown here with appropriately small teats, and train the piglets to use it as soon as possible.

very laborious, but after a few feeds you may find it possible to train piglets to drink from a small nipple feeder, enabling you to continue the frequency of feeds and vocal interaction without having to handle every pig.

Once all piglets (including runts) are able to drink from the nipple feeder, you will be able to progress to a more 'ad lib' style of feeding, where enough milk is provided for several hours, allowing you to get some sleep. By the end of a fortnight, you should gradually transition this to feeding warm milk twice a day in a (clean) feeder. At this stage piglets can also be provided with some solid creep feed. By four weeks of age they will be able to maintain their own temperature without a heat lamp (depending on the time of year).

TWO DAYS POST-FARROWING TO WEANING

Throughout lactation, the area should kept clean, dry, warm and draught free, with temperature fluctuations minimised. Piglets that look hollow should be examined and helped, and sow feed intake should be kept as high as possible. Throughout the lactation, flatten beds every day so that piglets don't roll into the dip and get squashed or injured by the sow. Old sows are more likely to flatten piglets so be on especially high alert, and consider stopping breeding from her if she becomes impossible to manage.

Sows' milk is deficient in iron, meaning that piglets are prone to developing anaemia, as explained in Chapter 9, 'Diseases of Pigs'. Outdoors, piglets can often get the iron they need from soil, but if they are indoors they may need supplementary iron. Sods (clumps of good quality soil) can be placed in the pen for both mum and piglets to enjoy rooting around in and from which they will gain the benefit of some iron. Alternatively, or in addition to this, an injection of iron can be given to the piglets at three to five days old, under supervision from your vet.

With regard to sexing the piglets, don't look at their tummies as boys and girls look very similar at that age due to the umbilicus. Instead, look under the tail: boys are smooth with just an anus, whereas girls have a triangular clitoris which, at that age, is often light-coloured or pink, even in black pigs.

If castration is completed by a vet, it's essential that feeding is observed over the next few days to ensure that piglets aren't in pain.

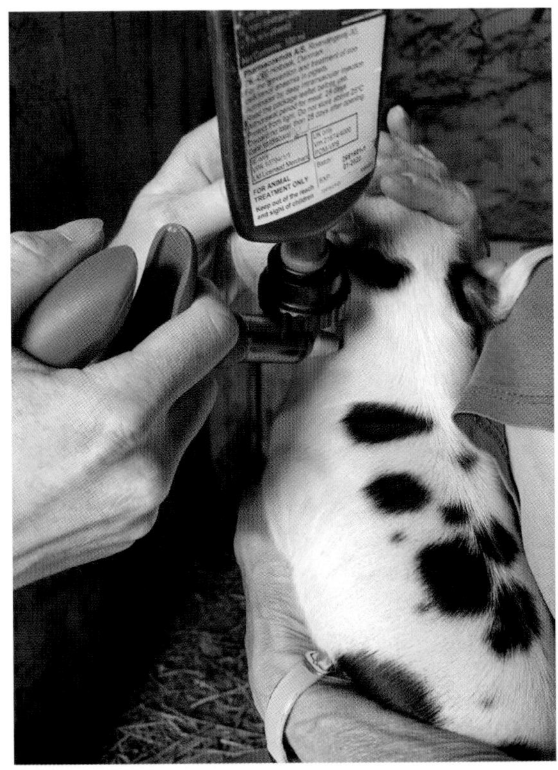

This piglet is receiving its iron injection at four days of age.

The left pig is a boy pig, as he has just an anus and sacs for testicles to descend into. The right pig is a girl pig as she has an anus and a vulva.

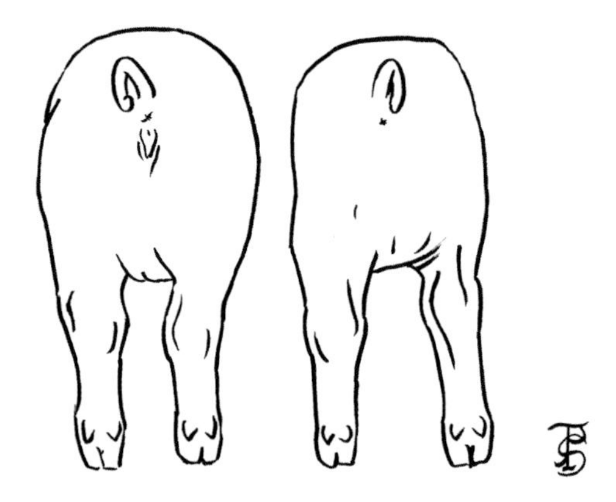

This time the female pig is on the left and the male pig is on the right.

Sue and Stephen find that offspring start to see the outside world at around day two or three. They keep mum and piglets indoors but with an outdoor space for the first two weeks, at which point they are moved outside to a paddock and a large arc.

By three days old the main risk of being squashed by mum is over as the piglets have now found their feet properly and can move quite quickly; however, piglets can be lost or, more commonly, injured by the sow at any age so you can never completely relax.

By law piglets must be provided with water from fourteen days of age, but Sue and Stephen find they start drinking from two or three days old. They need a shallow tray that they can reach easily (for example the lid of an IBC container or cat-litter tray). Mum's water, if open, should have a rock or two placed in it so that if a piglet climbs in it can get itself out and not drown.

Once the family moves into the paddock, Sue and Stephen find a separate piglet area very useful. It enables them to feed the piglets separately, have plenty of interaction with them to get them used to people without having to worry about a protective sow, and stops the sow drinking the piglets' water. They can also start feeding piglets hard feed, in a technique called 'creep feeding', which prepares their gut for hard food later on.

Piglets enjoying a drink of water from a shallow tray at three days old.

Interaction with the piglets is very important to make weaning easier and less stressful for all parties concerned.

Piglets will start to show an interest in mum's food after four or five days, and will get more and more interested as time goes on. Feed sow pencils so the food isn't too large for them, and spread the food out so they have a chance to get some before mum has it all.

Monitoring Piglet Disease

There are many signs of disease that piglets can show before weaning, from diarrhoea to coughing, which should be acted on accordingly alongside your vet. Injuries are also quite common, as piglets can be quite rough in play, and the sow can catch them

Piglets love to play boisterous games, which can sometimes end in injury!

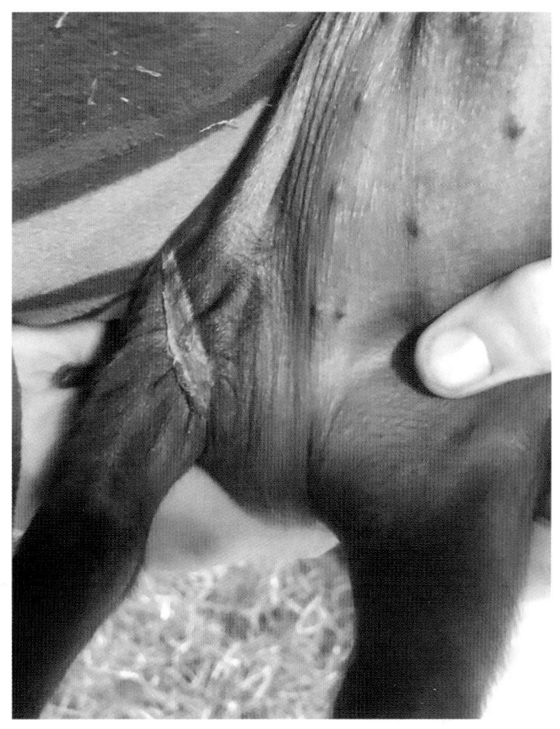

An injury such as this definitely requires a conversation with the vet! Luckily this piglet never became lame and the wound healed well under veterinary management.

A haematoma such as this should never be lanced by a keeper.

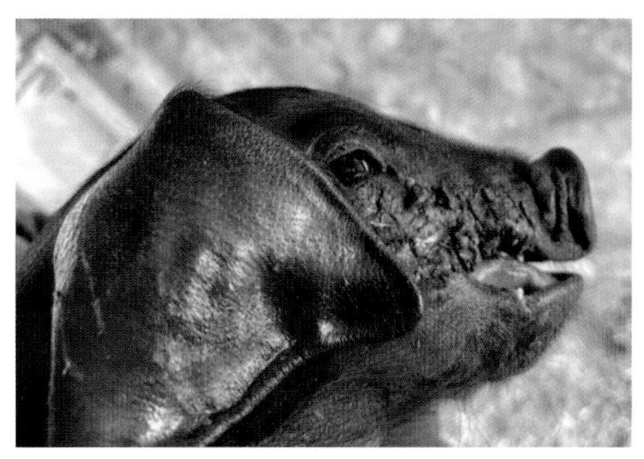

This piglet has a facial injury from competition at the teat. This needs to be promptly acted upon alongside your vet to ensure that it doesn't become infected.

Sue and Stephen carefully applied colloidal silver (under veterinary direction) to the injury, which allowed it to heal without problem.

lying down. Differentiating injuries from joint ill can be tricky and your vet must be consulted for advice on this. Piglets can also cause facial injury to litter mates from competition at the teat.

They can sometimes show signs of bleeding under the skin after an injury, which forms a soft, sometimes large swelling, known as a haematoma. These should never be lanced by the keeper (which would be illegal as it is an act of veterinary surgery), and are generally best left anyway. If you are worried, please contact your vet.

WEANING

Most smallholders will wean much later than commercial holdings, at between six to eight weeks. Note that twenty-eight days is the minimum age that piglets can be weaned legally, and forty days is the minimum for organic producers. Pigs will wean naturally at fifty-six to seventy days.[4]

Within this period, time of weaning must depend on the sow. Some sows will lose too much condition and we say that she is 'milking off her back'. Consider weaning earlier if her body condition score drops below 2.5. If a sow keeps her condition well and is more attached to her offspring, you can wait a bit longer. The behaviour of the sow and litter will tell you when they are ready to be independent from each other.

Piglets weaning at less than forty days will need more warmth and care. Remember that piglets should be eating a good amount of hard food by weaning to ensure that they cope with this stressful change.

Ideally, remove the sow from the piglets. That just isn't possible in Sue and Stephen's set-up, but they still find that weaning generally goes remarkably smoothly with barely any stress to the pigs. They wait until feeding time, then run the piglets up into the yard where they were born. Whilst the piglets are eating, and are therefore quiet, they walk mum over to a further paddock where she can't see her piglets and won't easily hear them. If done at evening feed, by the morning everyone is happy.

FROM WEANING TO SELLING

Different producers will decide either to sell weaners straightaway, or will wait for a week after weaning before selling. Either

Keep a close eye on the udder after weaning for a mastitis such as this, which could stop this teat being able to produce milk for the sow's next litter.

way, boys and girls should be separated very shortly from the point of weaning, depending on when you wean. They can breed much earlier than you would think, and don't mind mating with a sibling. This is especially the case if new pigs are added to the group, for example if two litters are weaned together.

Sue and Stephen keep their weaners in for a week both to train them and to ensure they have coped well with the transition to independence. At the end of the week they will tag and notch them if they plan to register any, and then the piglets are turned out into a paddock with plenty of forage.

Now is the opportunity to spend plenty of time with the new weaners. Call them whenever you feed them, give them treats, just sit in the pen and let them come to you. You will very soon have weaners that come to call and love being around people, which reaps dividends whether you are selling them or growing them on yourself. If they nibble or root on you, stand up and give your love to a different piglet.

Keep a close eye on the sow's udder to ensure that her milk dries up well. Call your vet if you see any signs of udder inflammation (hot, red, swollen).

BIRTH NOTIFICATION

All pedigree piglets should be birth notified with the BPA (or the British Lop Society or British Kunekune Pig Society, if appropriate). This can be done from three weeks onwards, and they ask that piglets be tagged and notched at this stage. When you add the piglets on the BPA Grassroots website you should list all the boys first followed by the girls, and each piglet will then be assigned a number. This is simply the number of piglets you have bred, so if your first litter was of ten piglets they will be numbered one to ten, and your next litter will start with eleven. This number is preceded by your three-letter BPA herd designation, so ABC1 for your first piglet.

REGISTRATION

Unlike many farm animals, pigs are only allowed to be registered if they meet the breed standard. In some breeds, such as the Oxford Sandy and Black (OSB), this involves an external assessment, whilst other breed clubs trust the breeder to make that call. This means that you need a good knowledge of the breed standards and how to assess a pig.

Breed standards are published by the BPA and can be found on their website.

It could be very tempting to register every piglet that conforms to Section A of the standard, as registered pigs can be sold for three times the price of growers, but as a breeder you have a responsibility to try to improve the breed. Pigs don't have to be perfect for registration, but you should certainly aim to have the majority of points mentioned in Section B. It is also easy to get fixated on finding the perfect underline whilst ignoring other conformational faults - it is no good a gilt having a great underline if her back legs are unable to support a boar! A show pig will need a perfect underline, but if the rest of the conformation is sound then some unevenness or an extra teat towards the back might be overlooked in a breeding rather than show animal.

Marking and Naming of Pedigree Pigs

Registered pedigree pigs are required to have two forms of identification, one of which will be a tag. The vast majority have one tag and either notches or a tattoo depending on the breed, and this is what is required for showing. If you have no intention of either showing yourself, or of selling your piglets to other homes where they might be shown, you can elect via the BPA to use two tags. For obvious reasons only white pigs are tattooed.

The tag should have UK followed by your herd number on one side, and the pig's BPA number with your herd designation on the other. A small possible confusion is that although the tag and ear number is derived from the number of pigs you have bred, the pedigree name is your herd name followed by the sow line (for gilts) or the boar line (for boars), and a number that is derived not from the number of pigs you have bred, but from the number of that line that you have registered. The first pig you breed of any line is unnumbered, then the next is two, and so on.

Notching is the removal of small, triangular pieces of ear in a pattern that translates into the pig's number.

The 'code' can be found on the BPA website, and also in printed herdbooks, but that for Saddlebacks is different to all the other coloured breeds. At first it seems a bewilderingly complicated system, but with practice and familiarity, reading the numbers from a pig's ear becomes quite straightforward. Notchers can easily be bought online.

SELLING WEANERS

Weaners, both breeders and growers, can be advertised via the breed society, on Sell My Livestock or through word of mouth.

We would advise 'vetting' prospective owners in some way to ensure that your pigs are going to good homes. Never, of course, sell a single pig unless it is going to a home which already has pigs. When selling registered breeding pigs you will need to ensure that the new owners are members of the BPA; if they are not, then although you will be able to transfer the registration to them, they will be unable to birth notify or register any progeny, meaning that that pig's offspring will, in effect, be lost to the breed. It would also be worth encouraging them to join the breed society.

When you sell you will need to set up the movement order on eAML2, then print off a copy and sign it, ready to give to the new owner. Make sure you put the correct identification for the weaners on the form, whether permanent or temporary, and if temporary, ensure you mark them as described to keep the new owner legal whilst transporting them home. Registrations can only be done by the breeder, so any breeders you sell must be registered on Grassroots via the BPA website before leaving you.

CHAPTER 14

PIG TO PORK

This chapter is intended to assist producers taking pigs from weaning to the finished product, with practical hints and tips. The wording is aimed at those collecting their first weaners, as we would like producers to start by raising weaners for slaughter before moving on to the much more complicated endeavour of breeding. We won't repeat material found elsewhere, so reading other chapters will also be necessary. We also encourage you to learn by doing, by completing Sue and Stephen's 'Pig to Pork' course, which this chapter is aptly named after. We promise that you will be Large Black pig enthusiasts by the end of the day! Sue and Stephen's course also contains much more information on selling your pork, so is an excellent progression from this chapter.

FROM WEANERS TO FINISHED PIGS

Preparations before Collecting your Weaners

You should have now confirmed a collection date of two or three male pigs from a local breeder who is happy to mentor you along your journey. You'll be keeping your pigs over the summer, all legal necessities will be in place, and you will have constructed suitable housing for them. A quarter-acre paddock can suit two weaners up to finishing weight, and a 4ft × 5ft arc will house them nicely.

You will have registered with a livestock vet, and you will have ensured that you have a suitable abattoir within reach. You will also have ensured that the breeder will worm the pigs before they arrive, with an avermectin-based wormer twice, two weeks apart before they arrive. A day or so before arrival you will need to give the breeder your address and CPH so that they can create the movement order on EAML2. You will receive an email once it is done, with a link to click to confirm the move once they have arrived.

Collecting your Weaners

Whilst with the breeder, make absolutely sure that you are taking pigs of the same sex: if they are not, they will breed together. Pigs under twelve months of age can be moved on a temporary mark made with stock marker, but that mark (red dot, yellow stripe) must be detailed on the movement order. Make sure that the breeder gives you a paper copy of the order to carry with you, and that the tags or marks on the pigs match those on the order.

When buying weaners, you don't want non-pedigree pigs to come to you with an ear tag in because they will need to be marked with *your* herd mark when they go to slaughter. The situation is different for pedigree pigs, which are under derogation to move to different farms or shows on their original herd mark – but they will still need to carry *your* herd mark when they go to the abattoir.

Weaners can be collected in the back of an estate car or pick-up (with canopy), perhaps in a large dog crate with straw bedding. Remember that you need a contingency plan for the weaners if you break down, so have with you water and feed and ways to give them to the pigs, just in case.

On Arrival

Let the weaners out into the paddock and leave them to explore their new home. There is no need to confine them to their arc for a couple of days as some suggest: they will soon find it and make it their own. They will not long have left mum and will never have been moved before, so this is a big and stressful day for them. Fresh grass to eat in the paddock will go a long way towards distracting them, but it would probably be best to leave them to settle until the next day before starting to get to know them.

Suitable piglet transport for weaners from a breeder.

These new weaners have arrived to an abundance of lovely fresh grass.

You should have some food from the breeder, so mix that with your food, and feed at least twice a day, preferably at roughly the same time, as pigs enjoy having a routine. Once you have left them to settle in, confirm the movement order, fill in your herd register, and don't forget to clean your transport thoroughly.

Once you have acquired your pigs, make sure that they are transferred to you on the BPA Grassroots database, which will enable you to sell the meat with Pedigree Pork certificates (if you are a member of the BPA).

Care of the Growing Weaners

Get your pigs well handled and trailer trained. You will need to make sure that you have a suitable trailer in which to send them to the abattoir later.

Many people will tell you not to name weaners so you don't become attached to them. Sue and Stephen do name their weaners so they can learn their names, which they consider makes them easier to handle. They use a general call for all pigs, plus individual names. If you use these whenever you feed or interact with your weaners, they will very soon come to call, which makes moving them, or retrieving them after an escape, a virtually stress-free process.

If you need to move them either for loading when they go, or to move paddocks, Sue and Stephen find having two people is best, although the second is often not needed. The person who normally feeds can call the pigs whilst carrying a trug of food, and they will generally happily trot along after it. The second person follows up behind with a board to steer and encourage any stragglers.

Sue moving weaners - yes, it really can be that simple!

In addition to twice daily feeding, it is essential to check the water supply and top up the wallows.

Pigs turn over the ground very effectively, so do a weekly check of the paddock to ensure they haven't dug up anything that could harm them, such as sharp pieces of old farm machinery.

Check pig drinkers daily for cleanliness, and empty and clean them out if they are dirty.

Peer into the arc daily to check the pigs still have sufficient bedding, that they haven't walked too much mud in, and that they haven't damaged the interior of the arc.

The Finished Pig

Most commonly, smallholders will send pigs as 'porkers' that will produce pork. Porkers should be sent at about 60kg, which is often at just over six months of age, but again this will depend hugely on the breed. Cutters (that produce larger joints) should be 75 to 85kg. Baconers (that produce bacon) should be 80–100kg. Some abattoirs will also have a maximum weight, so be sure to check this before your pigs go over that! Dead weight is about 72 per cent of live weight.[1]

Speak to your breeder about how quickly you can expect the pigs to grow and when to send them. Sue and Stephen find that their pigs grow at around 10kg per month, but this does vary hugely even between litters. To judge

This pig is off to slaughter next week.

This formula[1] can be useful to estimate the weight of a pig using body measurements when you don't have weigh scales.

$$\frac{\text{Heart girth (")} \times \text{heart girth (")} \times \text{length (")}}{400} = \text{Weight of pigs in lbs (1lb = 0.45kg)}$$

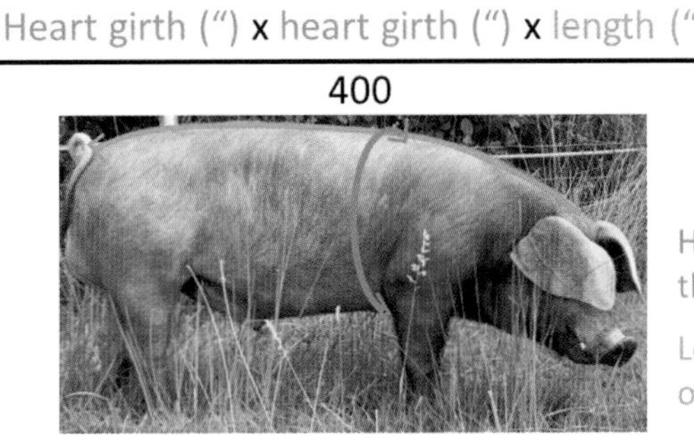

Heart girth = measurement around the body, just behind the front legs.

Length = measurement from the base of the tail to the base of the ears.

weight, weigh tapes can be very useful, or use the formula[2] depicted on the previous page.

You can also find this formula in a calculator form online from Phin Hall at https://www.phinhall.net/online-pig-weight-calculator/

SENDING THE FINISHED PIGS TO THE ABATTOIR

Depending on the abattoir, you may need to book in your pigs well in advance (potentially three months), so you must check this during initial conversations with an abattoir.

Completing a Movement Order

Once you have your pigs booked in at the abattoir, you will need to complete a movement order on eAML2. This can be done at any time in advance. You will have to print off a copy of the PDF you will be emailed, sign it, and hand it to the abattoir when you take your pigs.

Since January 2021 a new section has appeared on the eAML2 asking whether you have an 'up-to-date quarterly veterinary statement.' This is a post-Brexit addition, which only applies if your pork is leaving the UK to be sold in Europe. For the vast majority this will not be the case, so you can tick 'no'. If you are planning on selling abroad, please contact your vet for the required paperwork.

Before sending pigs, double-check that the abattoir is happy to take 'non-controlled' pigs: these require a Trichinella test, which is a legal requirement.

Preparing the Pigs

When pigs go to the abattoir, they will need to be identified with your herd mark. Speak to the abattoir about their requirements, but this is commonly done with a slap mark or an ear tag, especially for a coloured pig. They must also be clean and 'starved'.

It is so important that this process is stress free. Firstly, a good death is important to every smallholder we know. Secondly, stress causes meat to go bad.

Ear tagging: The normal slaughter tags are small metal ones that fold over the edge of the ear. The disadvantage of tags is that unless you have the head back, you will have no way to confirm that the carcass is that of your pig. You will also need to purchase tags for each pig you finish, and although each tag only costs around 50 pence, the postage tends to be expensive, so the cost mounts rapidly. Go to Chapter 15: 'Common Procedures' to learn how to ear tag.

Slap marking: A slap mark is a sort of tattoo of the holding herd mark on both shoulders of each pig. The marker leaves a tattoo that is visible on the meat, so you can confirm you have the correct carcass. The initial expense is more than tagging (around £70), but once you have purchased it you only need ink thereafter.

Cleanliness: Pigs should be reasonably clean, and Sue and Stephen do this by bringing them in the day before. They either bath them, or just leave the mud to dry off then brush them clean, which the pigs love. A flexible curry comb is perfect for this.

Not fed: The pigs' stomachs should be all but empty, so don't feed them on the morning they go. You will inevitably have to be at the abattoir very early, and they will probably be dead before their normal breakfast time. A bit of food to lure them and a bit of carrot or apple as a treat is fine.

Loading and Travel to the Abattoir

Please don't load pigs the night before without water. This is illegal and completely unnecessary. If you have trailer trained your pigs this does not need to be a stressful event.

Pigs love their comfort and are very curious, and Sue and Stephen make this work to their advantage. They put the trailer in the entrance

A flexible horse curry comb can be used to brush pigs clean.

Sue and Stephen's well-thought-out set-up for loading for the abattoir.

to their little yard as shown, and ensure they have water in the yard. They bed the trailer with straw, and put a rubber mat on the ramp to make it more inviting. They then close the stable door so the pigs can't get into their normal shelter. With no other access to shelter and bedding, the pigs go to sleep in the trailer, so in the morning Sue and Stephen simply tip-toe into the yard and close the trailer door.

In a paddock this can be emulated by putting the trailer in the field gate and blocking off the arc, or putting the trailer at the door of an unbedded stable or other secure area would also work.

Pack your car the night before. You need equipment in case you break down or have an accident, so food and water for the pigs and a pig board are essential.

TO AND FROM THE ABATTOIR

Go to visit the abattoir beforehand so you know what to expect, and you can make sure you are happy with how the pigs will be treated. There is a list of abattoirs that take pigs on the eAML2 website, and word of mouth is always a good indicator, although isn't necessarily reliable. Handling should be quiet, patient, gentle and sympathetic, lairages should be well bedded, everything should be clean, and pigs should move through the process calmly.

Ensure that you are clear about where to take the pigs, where to give in your movement licence, where to collect the meat and pay, as they all tend to be different places. This will also give you a chance to practise your trailer reversing at home once you know what sort of approach there is to the lairage. The morning you take your first pigs will be stressful and probably very early, so being happy with how to find it and where to go when you arrive will pay dividends.

What to Expect When You Arrive

You will be told to arrive in a bracket of, say, between 06:00 and 08:00. Although it might be tempting to go for the end of the slot, the abattoir will be busier then, so it is worth arriving earlier, especially for your first time. Sue and Stephen tend to turn up shortly before 07:00. You might have to queue, but when it comes to your turn, reverse your trailer up to the lairage gate. There will be a

Sue and Stephen's pigs, Stumpy and Derek, making friends in the lairage.

stockperson waiting for you who will secure gates to either side of your trailer. When they are secure, open the trailer and lead your pigs out. The stockperson may put the pigs into the indicated pen, or you might be expected to. It is worth having a trug with a little food to lead them with.

Once they are secure, move your vehicle and trailer out of the way, and take your movement licence and cutting list to the office (you will know where this is from your recce). Depending on the abattoir, you will either leave them on the pile, or hand them over in person and talk through your cutting list.

When You Get Home

The following tasks must be completed quickly when you arrive home:

- Clear the bedding out of the trailer and wash it out.
- Record the movement in your herd register.
- If you are in the BPA, report the pigs as dead using the Grassroots system, and download your pedigree meat certificate/s.
- Go back to eAML2 and click on 'Confirm a Move'. You might well need to add a bit of information. This isn't always necessary but needs to be checked just in case.

Over the coming weeks think about what you want to do with your beautifully turned and fertilised area of ground. If you would like more pigs in the same paddock you will need to rest the paddock for several months. You can let it recover naturally or resow it with a pig rooting mix. These can contain stubble turnips, fodder rape and other piggy treats, as well as grass.

If you decide that keeping pigs isn't for you, you have the perfect spot for a vegetable garden.

THE END PRODUCT

Planning

Before sending pigs, have a general chat with the abattoir about cutting lists and sausage flavours. You will need to supply a cutting list for the abattoir, unless you plan to butcher the pigs yourself. The cutting list can be quite daunting to work out first time, but to give you an idea, see the box [opposite], which shows an example of one of Sue and Stephen's cutting lists (for porkers).

For porkers, there are always compromises to be made. For example, you might want the head back for brawn but the cheeks generally go into sausage, so by keeping the head you will considerably reduce the amount of sausage you get. You have to sacrifice something to get sausage, and for Sue and Stephen it is the shoulder and head. Another thing to bear in mind is that the sausage mix comes in large and fixed quantities, so that generally you can have only one flavour per half pig. Again, check with your abattoir.

If you plan to sell the pork, you must ask for it to be vacuum packed by the butcher unless you have special facilities.

EXAMPLE CUTTING LISTS FOR TWO PIGS

Cutting list for a large pig:

This larger 70kg pig gave Sue and Stephen twenty-two packs of four sausages. These were large sausages, so two per serving.

Cutting list for one half:

- Shoulder as sausage
- Joints sized for four (this means the leg will be cut into two)
- Chump as chump (that is, as a joint, not sausage)
- Belly as joints
- Loin as chops
- Trim as pork and apple sausage (thick) packed as fours

Cutting list for the other half:

- Shoulder as sausage
- Joints sized for two (leg will be cut into three)
- Chump as chump
- Belly as strips (packed in fours)
- Loin as chops plus two small joints
- Trim as pork and apple sausage (thick) packed as fours

Cutting list for a smaller pig:

- Shoulder as sausage
- Joints sized for two
- Chump as chump
- Belly as strips (packed in fours)
- Loin as chops plus two small joints
- Trim as pork sausage (thick) packed as fours

You can request that all should be vac-packed, and that you would also like the trotters and pluck to be returned to you if possible.

Collecting your Pork

You will be told when to collect your pork, and as your pigs will be 'non-controlled' if you keep them outside (this is a question on eAML2 and refers to their housing, not their temperaments!) they will need to be tested for trichinella, which adds a day on top of the standard butchery time, taking around four days.

Some abattoirs will put the meat in bags, others in boxes, but all will expect your car to be reasonably clean. If you are selling the meat, use insulated boxes to maintain it at a safe temperature.

If you have asked for the 'pluck' back, remember that it might not pass the vet. The other thing to bear in mind is that if you have dark-haired pigs, you might not get the trotters back as it can be difficult to remove all the bristles.

Storing your Pork

Once you get your pork home you need to sort it immediately, and you will need to have an area set up ready with a table and chair close to the freezer. If any customers are collecting pork on the day it needs to be put into boxes and labelled, then left in a suitable cool place for them to collect, and all pork for sale that

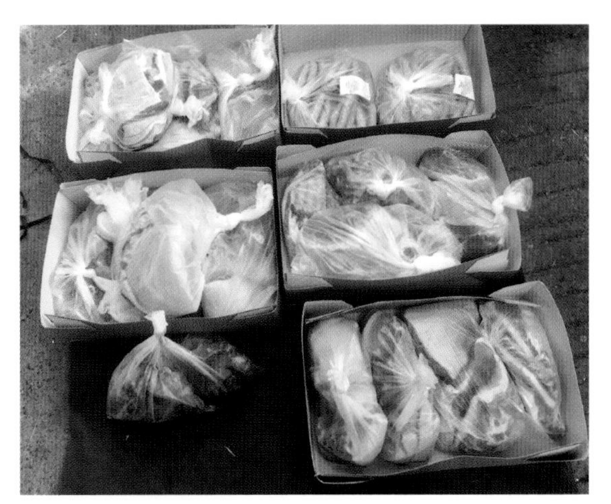

This is the meat that two small (55kg) pigs produce.

is being frozen needs to be labelled before going into the freezer.

As a minimum you need your company logo, a 'best before' date, and storage information. Consult the 'assurance schemes' and 'society memberships' sections of this book for other labels that you might want to display. There are many legalities involved in the labelling of meat, and you must speak to your Environmental Health Officer regarding label requirements, and make sure that you meet all legal requirements before doing so.

A butchered pig produces a surprisingly large amount of meat, and you will need room to freeze it. Sue and Stephen have a 465ltr commercial chest freezer, and this will just hold three butchered pigs. However, they take no more than two pigs at a time as this is a lot of meat to freeze in one go - it takes two or three days to freeze down properly. Remember this if you are sending frozen pork by courier - you will need to allow a few days between collecting the meat and sending it out.

The very last thing you need to do is reward yourself for having worked for four months or more to produce wonderful, high welfare, delicious pork, by having your first home-produced pork dinner.

AN ALTERNATIVE END

If you don't manage to send your pigs to slaughter, don't worry. There are many pigs intended for a producing purpose whose owners form such a bond with them that they are kept on as pasture ornaments. This is not failure: you have just assigned an animal a different purpose in life, and that is to bring you joy of a different form.

This lovely pig was hand reared as a piglet so was kept as a pet; however, it is now around 200kg!

However, you must consider that pigs live for ten to fifteen years. Boars can grow to 500kg and can be dangerous, such that farm-animal sanctuaries are struggling with how to take care of them. Due to not being bred to be kept as a pet, many will develop diseases such as arthritis, requiring long-term medication to ensure they maintain a decent quality of life. Therefore, when considering not sending your pigs to a more sausage-shaped end, please consider how you will maintain their care for the rest of their lives.

As tempting as it may be, you must not keep these pigs for breeding. They have already been determined as unsuitable for breeding by someone experienced at this choice. If you have purchased males and females, they are also likely to be either brothers and sisters, or very closely genetically related so must not breed together. This is likely to cause problems from birth with the piglets.

CHAPTER 15

COMMON PROCEDURES

Only the procedures deemed relevant to a smallholding will be discussed here. For example, teeth clipping should not be necessary due to smaller litter sizes in rare breeds, leading to less competition at the sow's teats. Likewise, tail docking should not be necessary on smallholdings. These procedures are illegal without a veterinary derogation.[1]

IDENTIFYING PIGS

Pigs are required to be identified for legislative and pedigree pig reasons. All pedigree pigs registered with the BPA require two methods of identification. The primary identification will be a tag, and the secondary method generally takes the form of either a tattoo or notches. Although it is possible to opt out and use two tags, this limits the market for the pigs as they cannot be shown. Pigs with pink ears are tattooed, but this obviously isn't possible in black-eared pigs, so instead the ears are notched.

To learn how to complete these procedures, we would recommend going to a fellow keeper to learn how to do this the first time.

EAR TAGGING

Ear tagging is commonly used for pigs that will be kept on a holding, such as for breeding or for pets, but can also be used as identification to send a pig to slaughter. Suitable tags will depend on whether you are tagging in order to register a pig as pedigree, or to send to an abattoir. The latter will just need 'UK', followed by your herd mark for identification, and will need to be printed, not handwritten. Please note that it is illegal to remove an ear tag once applied.

TYPES OF EAR TAG

Check with the abattoir as to the type of ear tag that they accept, which may need to be metal due to the high temperature of the scalding tank. Don't use tags that are closed at one end on piglets as these can cause injury when the pig's ear grows.

Correct positioning for ear tagging a pig using the side of a trailer.

Ear tagging is best completed while piglets are still small enough to be picked up. If you do need to tag older pigs, you can use the side of a trailer with a towel over the gate so the pig can't slip. This is a two-person job and both people should be wearing ear defence. Alternatively, you can do this with much larger pigs with their head in a feed bucket.

Procedure

- Prepare any equipment you might need, and load your clean ear tag into a clean and disinfected pair of taggers. Ensure that the applicator is designed to fit the tags. Make sure that the tagger and tags are suitable for the size of the pig.
- Ensuring safety around the sow, get the piglet in position and wait for it to settle. Hold the tagger so that the male side of the tag (with the stem) goes on the outside of the ear.
- Tag centrally on the ear so that the tag won't be hanging off an edge, but so that you don't hit the vein that runs down the ear. If using flag (triangular) tags, position the sharp point at the middle point of the ear. Press as firmly as you can with the tagger while the holder holds the pig very still.
- Place the pig gently on the floor.

Press firmly with the tagger.

Clean ear taggers should be prepared, as shown.

SLAP MARKING

A 'slap mark' is a method of identification for pigs going to slaughter. The process leaves the holding's herd mark over each shoulder, which, once the carcass has been through the scalding tank, can be easily read for food traceability. It is important that each shoulder is marked as the carcass is cut in two.

The slap marker is a plate on a handle, with a slide-in piece with pins displaying your herd number. This is dipped into ink and then slapped on to the pig's shoulders.

Marks can fade over time, which can be a problem if pigs are slap marked too far in advance before going to slaughter, so it is best to do this a few days before they go to the abattoir. It is also best applied to clean pigs, so doing this after their pre-slaughter clean generally works well. Be sure to take note of abattoir feedback on slap marks and adjust your technique accordingly.

Stand on the opposite side of the pig as Sue is doing and slap each shoulder with one firm motion.

SLAP MARKING FOR BLACK PIGS

Generally slap marking is accepted as identification for black pigs, as the mark is only required to be read after the pig has been through the scalding tank, which removes much of the skin pigment – but do check with your abattoir just in case.

Procedure

- Prepare the equipment by ensuring the needles of the slap marker are clean, well maintained and that all are straight. Every needle must be complete. Make sure that the plate is held securely in place, and check that your ink pad is serviceable and filled with ink.
- Dip the plate into the ink so that all needles come into contact with it.
- Stand next to the pig with its head in a bucket of feed, on the opposite side to the side that you are going to slap. Clean the area over the shoulders where you are about to slap.
- Slap the slap marker once on to each shoulder with one motion per side. Do *not* slap again, even if you are not happy with the placement of the mark.
- Refill the ink pad and recheck the equipment between every pig.
- Clean the slap marker thoroughly between uses.

Remember to move round to the other side for the opposite slap mark.

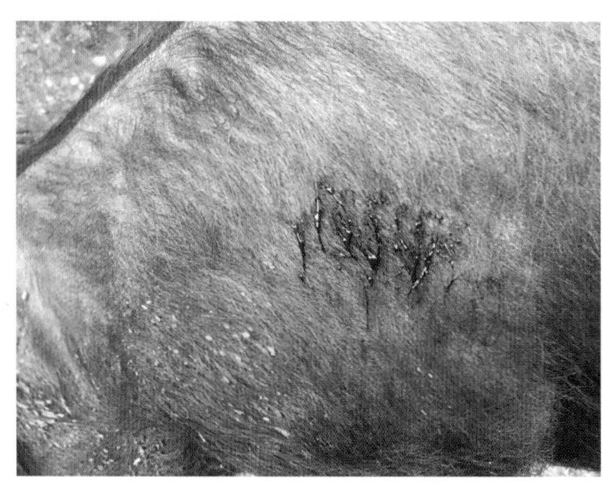

After marking, the slap mark may look quite unremarkable, but marks are far easier to see after the pig has been through the scalding tank.

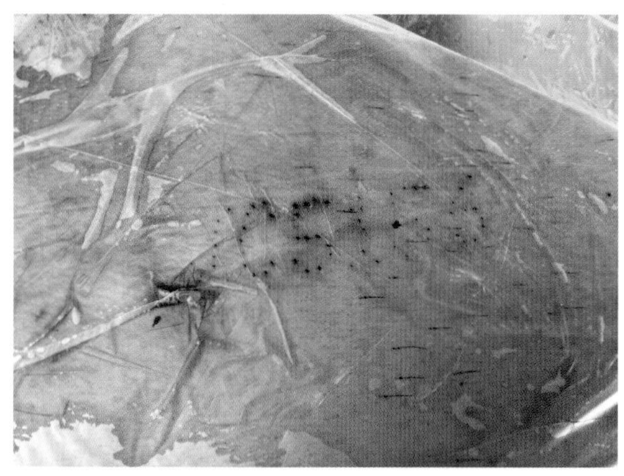

The slap-marked herd number JC5156 can be seen on this shoulder of pork.

Rehearse the motion before slapping as you only get one chance! The motion needs to be firm, but does not need to be hard or cruel. Only a tiny section of the needle needs to enter the skin. If done correctly the pig tends either to ignore it or just jump slightly.

TATTOOING

Pigs are best tattooed at about four weeks of age when they are small enough to be held. The pliers need to fit on to the ear, but you also want the tattoo to grow with the ear as the piglet grows. Tattooing requires two people, one to hold the piglet and one to tattoo, and ear defence must be worn.

Procedure

- With cleaned and disinfected equipment, insert the correct letters into the pliers. Check that the needles are straight and undamaged. Close the bracket so that they are secured.
- Test this on a piece of paper to check that the letters are the right way round.
- Put some tattoo paste on to the needles with a disinfected toothbrush, and smooth some paste on to the pig's ear.
- Holding the pig tightly, put the ear between the pliers so that the needles go into the top of the ear.
- Squeeze the pliers until you hear a little crunch.
- Release the pliers and rub tattoo paste into the area that you have just tattooed.
- Gently release the pig.

First insert the correct letters into the pliers.

Close the bracket so the letters are secured.

Smear some tattoo paste on the pig's ear and use a toothbrush to put it on to the needles.

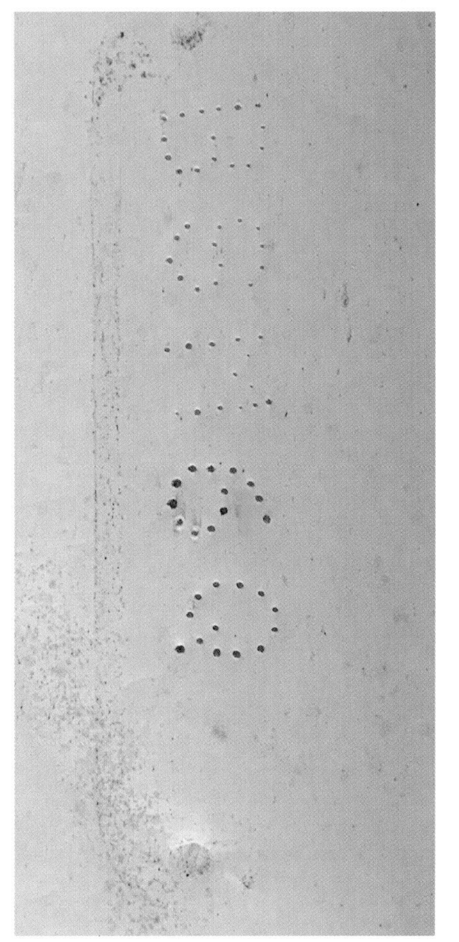

Test the letter order and orientation on a piece of paper.

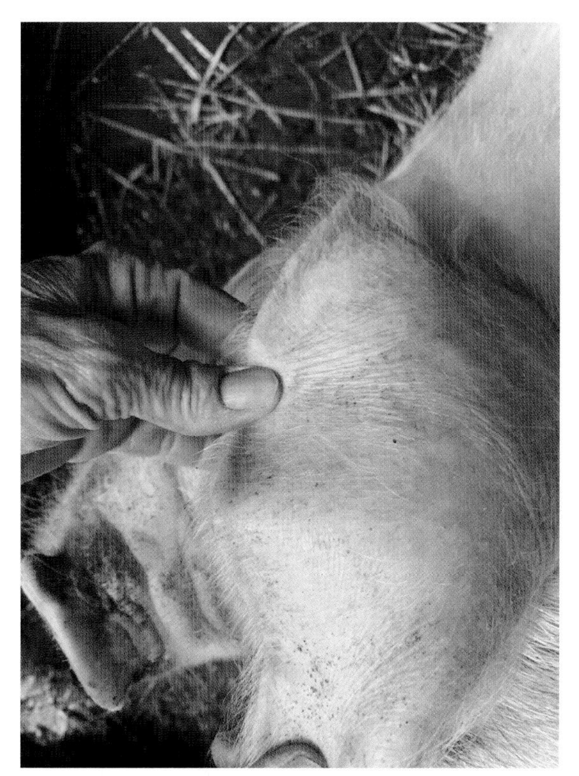

This tattoo is clear and has grown nicely with the pig.

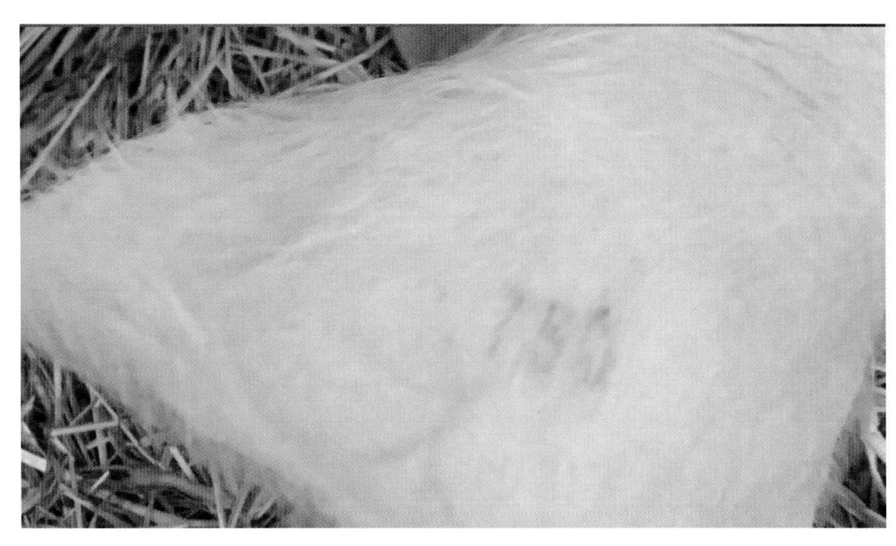

This tattoo is nice and clear but is quite small in comparison with the pig's ear.

NOTCHING

Notching involves using a specialised tool to clip out small pieces of the ears in a pattern that corresponds with the pig's number. The BPA asks that pigs are tagged and notched before being birth notified, the minimum age for which is three weeks, but we find that both are better done at around eight weeks when the ears are larger and more robust.

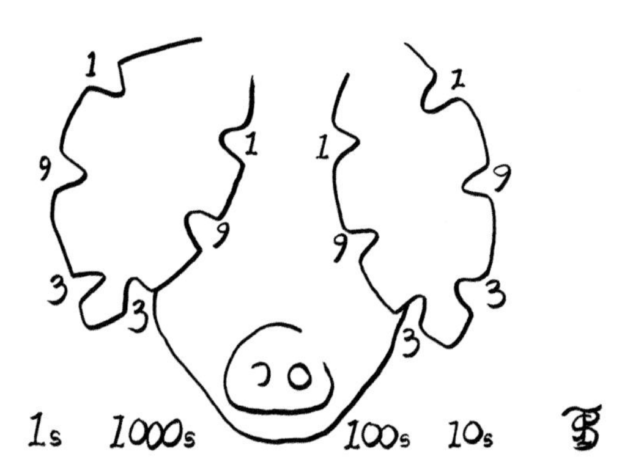

The ear-notching system for most breeds of pig, showing how the numbers 1-10 are represented by notches on the outside of the pig's right ear (left-hand side when shown head on), multiples of 1,000 on the inside of the right ear, 100s on the inside of the left ear and 10s on the outside of the left ear. The fewest notches possible should be used, so for example the number 7 would be indicated by two notches at the '3' position and one at the number '1', so 3 + 3 + 1 = 7.

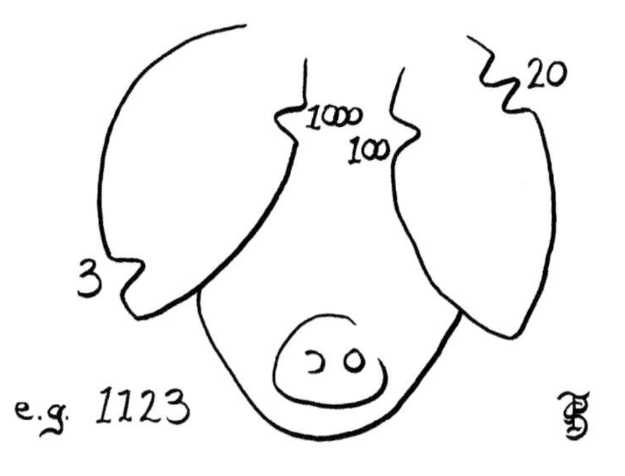

An example of the ear-notching system for most breeds of pig. In this case the pig's ID number is 1123, so there is a single notch at the '1' position on the inside of the pig's right ear indicating '1000'. A single notch at the '1' position on the inside of the left ear denotes '100', Two notches at the '1' position on the outside of the left ear means two lots of 10, that is 20, and a single notch at the '3' position on the outside of the right ear is '3'. 1000 + 100 + 20 + 3 = 1123.

Various notchers are available, but for pedigree pigs ones that remove a 'V'-shaped piece of ear should be bought. There may be

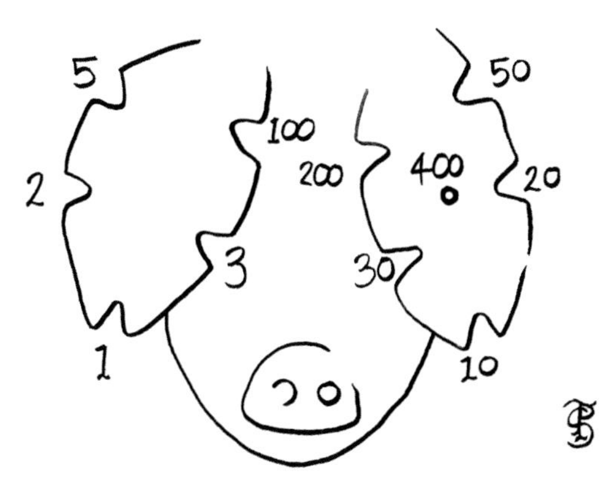

This diagram shows the ear-notching system for the British Saddleback. The principle is similar to other breeds but the numbers are not assigned in a decimal way, rather each notch has its own number, as shown with the addition of a single punched hole in the middle of the left ear representing '400'.

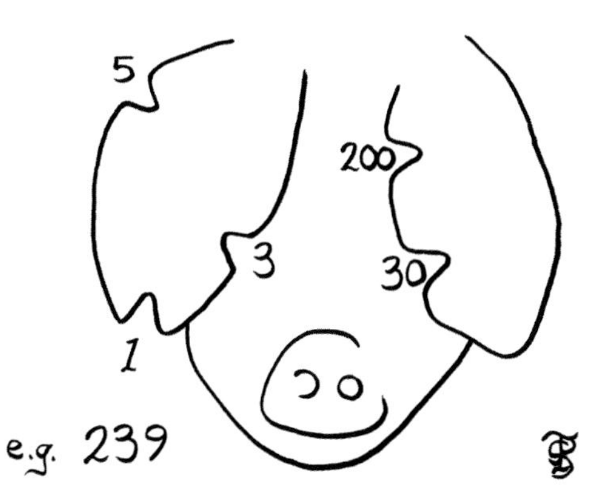

This Saddleback's ID number is 239. Individual notches at the '200' and '30' positions give us 230. To get the number 9 with the fewest notches, we place one at the '5' position, one at '3' and one at '1'. 200 + 30 + 5 + 3 + 1 = 239.

a slide that can be adjusted to control the size of the notch, and if so, this should be set to its smallest size (about 3-5mm) as the ears, and therefore the notches, will grow considerably. If there is not a slide, be very careful not to make the notches bigger than 5mm, and preferably slightly less.

Similarly to tattooing, notching also requires two people and ear defence.

Procedure

- Look up the notching system for your breed and draw a template of the notches that need to be made for each pig, as it is easy to get flustered when working with a screaming weaner, and mistakes cannot be rectified. Ensure that equipment is clean and disinfected.
- Pick up the weaner and hold it closely, waiting for it to settle. As long as pigs are

Correct equipment, cleaned and disinfected, ready to ear notch pigs.

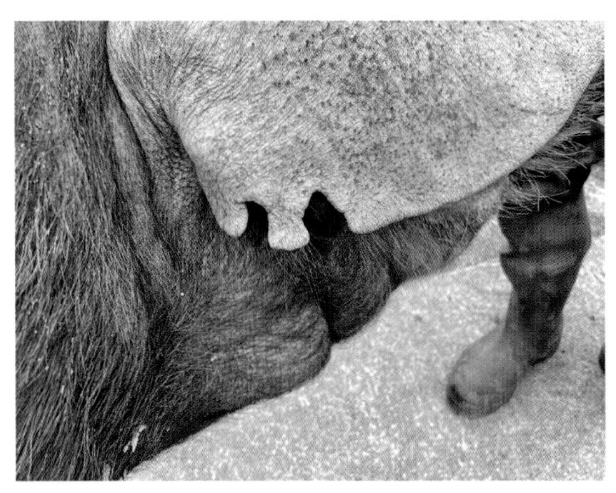

These notches were made too large, leaving the ear vulnerable to injury.

correctly restrained, they rarely show a reaction.

- Put the notcher in the correct place, ensuring that the ear is against the sliding guard, if present, and firmly close the handles. A small triangular piece of skin should be cleanly removed.
- Carry on until all the pig's notches are complete, checking the template frequently. Once they are done, use a spray of antiseptic on the fresh notches and release the pig.
- Clean the notchers between each pig.

The notches should heal quickly with no after effects, but keep an eye on the wounds for a few days just in case of infection.

DETUSKING

It is not essential to trim a boar's tusks, but there are three main reasons for doing so: to avoid injury, in the interests of human safety, and to avoid ingrowing tusks.

Pig injuries and behaviour: Boar's tusks can be real weapons, capable of inflicting serious injuries on other animals - particularly other boars or younger pigs. Once a boar becomes aware that he is capable of causing such damage, he may become unmanageably aggressive.

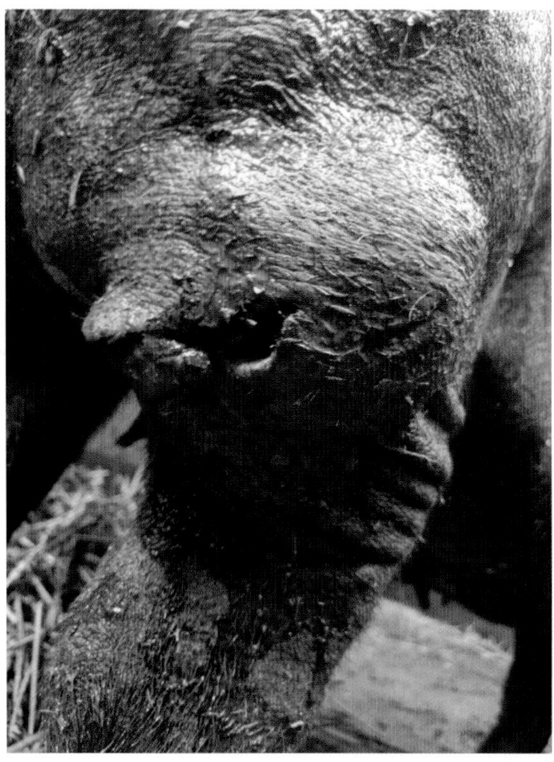

This serious injury was from the tusk of a non-aggressive boar.

Bran's tusks are quite the impressive instruments!

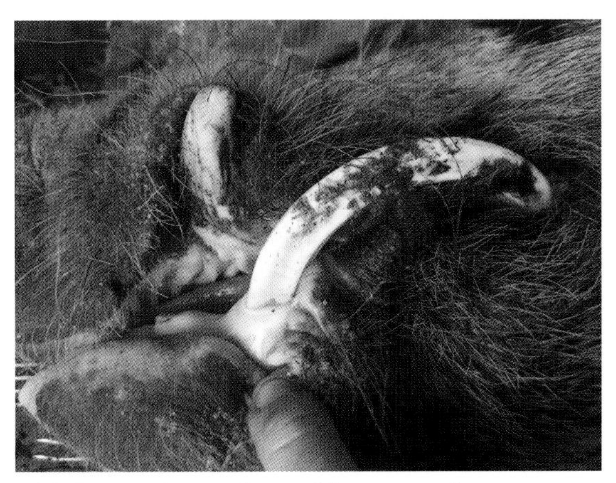

This ingrowing tusk would cause serious problems if not removed.

Human safety: Handling an adult boar can be dangerous regardless, but tusks increase the risk of injury even further.

Ingrowing tusks: Occasionally tusks can begin to grow into gums or cheeks, causing uncomfortable sores and areas of skin inflammation. If left unmanaged these lesions will affect the boar's appetite, and in extreme circumstances may make eating impossible.

RESTRICTIONS ON BOARS WITH UNTRIMMED TUSKS

Boars are not allowed to be shown with untrimmed tusks, and the convention is that boars with untrimmed tusks should not be hired out due to the danger to sows.

Tusks are the pig's canine teeth but, unlike ours, they keep growing throughout life. The tusks of male pigs generally grow faster than the tusks of female teeth, meaning that the tusks of female pigs rarely present one of the above problems. Male pigs may need detusking up to every six months.

It is advisable that detusking is carried out by your vet. Firstly, the correct location of the cut is crucial. If tusks are cut too short and the pulp cavity is exposed, pigs can get a tooth root infection. Secondly, the vast majority of boars require experienced handling and/or sedation to tolerate the procedure.

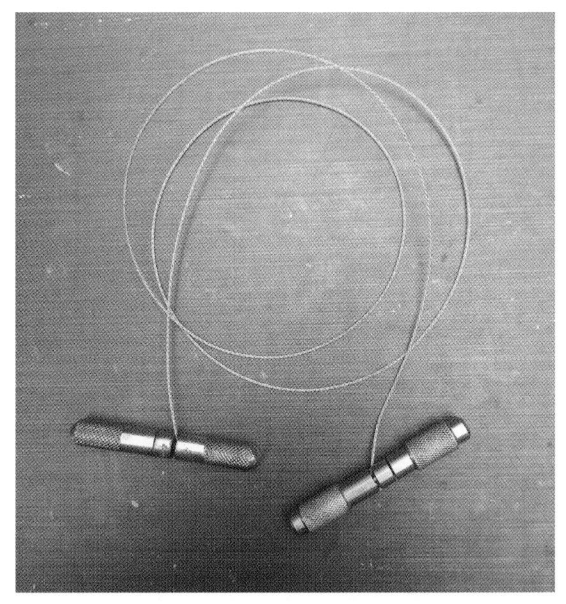

Dehorning wire is the only appropriate tool for removing tusks, along with a sensible method of restraint, either chemical or physical.

Dehorning wire should be used, which your vet should be familiar with. Bolt cutters, pliers or any kind of crushing tool are absolutely inappropriate because they will cause the tusk to shatter down below the gumline, leading to considerable pain and a high risk of tooth root infection and abscessation.

Procedure

Depending on the size and temperament of the boar and the facilities on the holding, this procedure can be completed in a small space with the pig loose, with the pig on a snare, or with sedation. To complete the procedure loose with an amenable boar, erect a small handling facility, or have the boar in a small pen. A narrow area will make the task far easier. To complete the procedure on a snare, consult Chapter 6, 'Handling'. If sedation might be required, be sure to discuss with your vet whether you should starve the pig beforehand, and their recommendations after the procedure in terms of warmth and water.

The wire should be hooked round the tooth and the handles held firmly. Then the tusk should be sawn above the gumline, perpendicular to the tusk growth. This area of the tusk does not have sensation and therefore the procedure is painless. Any lower than this risks exposing the tooth root cavity.[2]

Here, Bran is getting his tusks trimmed under sedation by Sue and Stephen's excellent vet, Kate, from Dyfed Farm Vets.

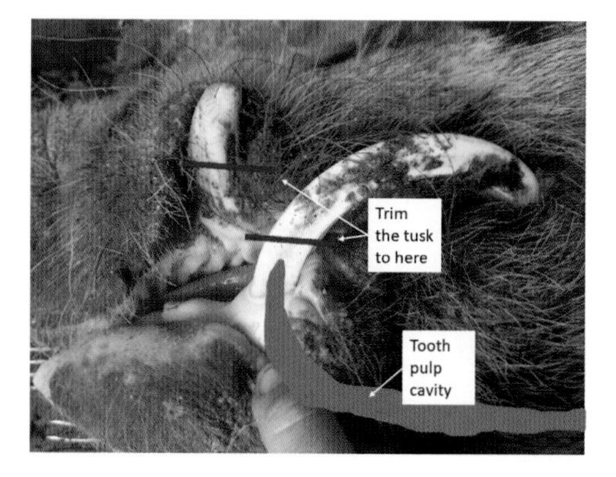

Trimming the tusks too low risks exposing the tooth pulp cavity.

FOOT TRIMMING

Foot trimming is unnecessary for the vast majority of pigs, but can be required where hooves become overgrown. This tends to be especially the case amongst pet pigs, in particular where conformation (body shape) is not optimal. The need for foot trimming can be reduced by keeping pigs on a hard standing for an hour or so a day, or walking them over concrete so that hooves can be worn down.

Foot trimming may also be necessary should a pig develop a foot ulcer. This is quite rare in pigs, but veterinary attention should be sought to deal with lameness that could be caused by a foot problem.

Preparation for Foot Trimming

Preparation for the potential need for foot trimming should start in pet pigs from a young age by getting pigs slowly used to rolling over for belly rubs. Over time some hoof trimmers can be introduced and eventually little snips can be taken off. An alternative method is to train a pig to lift its foot for food, which eventually progresses to foot trimming as well. Very similar methods can be used as for cooperative care in dogs,[3] ensuring that treats are suitable for pigs.

Alternatively your vet can sedate the pig so that it rolls over and allows the process. This is far preferable than stressing your pig in order to trim its feet, so where training is not achieved or is unsuccessful, your vet will be happy to assist. Also bear in mind that the claw will get harder as the pig ages so you may find that later in life you need some assistance.

Does My Pig Need a Foot Trim?

Feet should be trimmed if hooves are overgrown, meaning that foot functionality is impaired. Asking the following three questions can be helpful:

Is the pig walking on its toes? Pigs should walk on their toes, however when claws are overgrown, the weight starts to move to the back of the foot. This can then lead to joint or skin issues, when a foot trim is generally necessary.

Are the claws curving inwards? The claws should point straight ahead and not curve inwards.

Are the dewclaws below the hairline or touching the floor? The dewclaws are the two little claws at the back of the foot. These should not extend to below the hairline, nor should they touch the floor - however, do be aware that this can be a conformation issue rather than a foot-trimming issue in some pigs.

Procedure

Feet should be pared back so that the claws are not overgrown and the pig is encouraged to walk on the appropriate part of the foot. However, keepers must not over-trim as this could allow bacteria to enter the foot and would be painful for the pig. As soon as you start to feel softer tissue - stop. This is long before you cause any bleeding. If more hoof needs to be removed, revisit this in a few weeks' time to give the quick some time to recess.

Use a pair of hoof shears to take little slithers of hoof. Do not use hoof nippers as

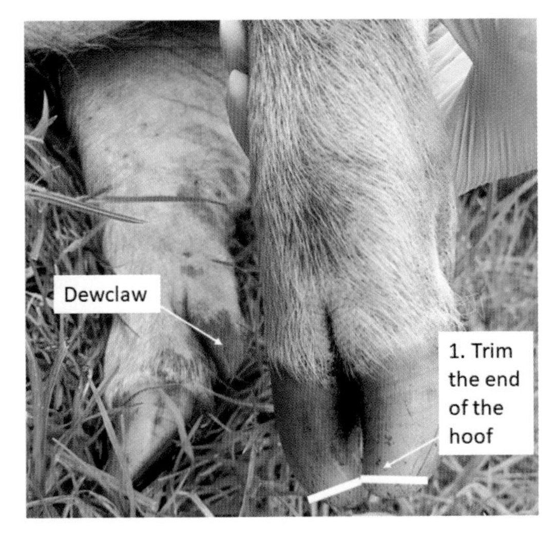

Start by trimming the end of the hoof to a normal length.

Then straighten up the walls.

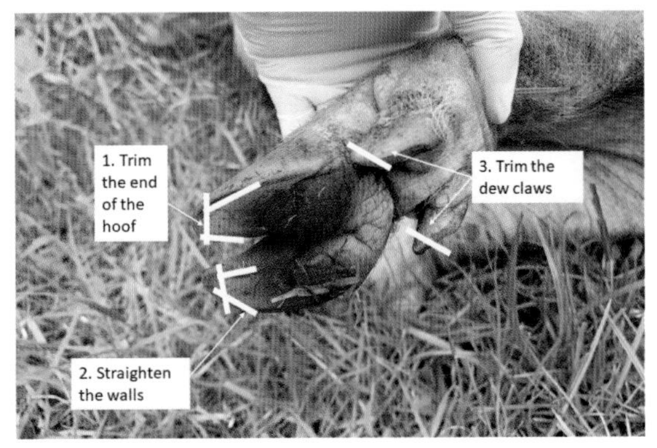

Finally, trim the dewclaws if required.

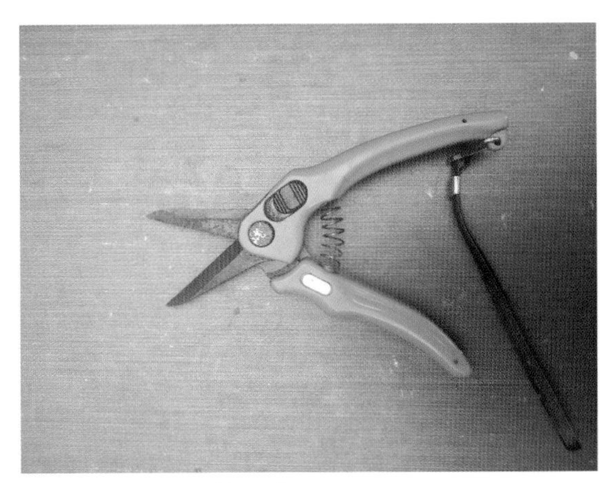

A suitable pair of hoof trimmers.

these are far less accurate. One claw of each foot will be a little larger than the other: keep it that way, and don't change the natural anatomy of the foot.

You can use a file at the end to tidy up the procedure, however the pig will do this itself naturally by walking on some concrete.

NOSE RINGING

Nose ringing involves driving a steel ring through the cartilage in a pig's nostrils so that it can no longer root the ground so that grass cover may be maintained. Legally this procedure can only be carried out on pigs living outside,[4] and boars cannot be nose ringed.

It is technically legal for this procedure to be carried out with no pain relief, by farmers or smallholders who are considered to be suitably trained - however, we really dislike this practice. In order to preserve some grass, a pig undergoes an acutely painful procedure, which then causes it pain every time it tries to exhibit natural rooting behaviour. We do not think this is ethically sound, and believe that if a piece of ground is not suitable for pigs, they should not be put on it. It should also be noted that pig feet also churn up ground, so grass cover may still not be maintained, even when noses are ringed.

We especially recommend that you do *not* do this yourself, but discuss it with your vet. With the pig on a snare, pliers are inserted into the nose holding the ring, which is driven through the cartilage.

A pig with a nose ring.

NEUTERING

Castration

Pigs are generally castrated for one of three reasons:

- To prevent inappropriate breeding, for example with siblings.
- To prevent boarish behaviour in pigs, such as mounting and aggression.
- To prevent 'boar taint'.

For male pet pigs, the prevention of inappropriate breeding and boarish behaviour are both excellent reasons to castrate. Therefore, early castration in male pet pigs is really important.

In non-pet pigs the reasons are far less clear. In a production setting, inappropriate breeding can be prevented by keeping male and female pigs apart.

Boar taint is the development of a smell or taste of excrement in pig meat, which can develop in boars from the point of puberty, and increases with age and weight. Luckily this is very rarely a problem on smallholdings. Commercially in the UK pigs are not castrated but are reared to much heavier weights than smallholder pigs, being slaughtered at around 120kg live weight. It is likely that the British public are less sensitive to boar taint than other parts of the world that do castrate pigs, but do not experience a problem even at heavier weights. Even smallholders who have sent boars to slaughter that have previously 'worked' (made sows pregnant) do not report a problem in the meat. Therefore it is inappropriate to castrate pigs that are being sent for slaughter for meat as it constitutes an unnecessary, painful surgery.

How is Castration Achieved?

Castration must only be carried out by a veterinary surgeon as an act of veterinary surgery. There is technically an exemption by which keepers can castrate their own pigs at less than one week of age,[5] however this is not allowed under Red Tractor due to the dubious ethics of this practice. It is our very strong belief that this painful procedure is not suitable for keepers to complete, especially as anaesthetic cannot legally be provided by a lay person.

The method by which a vet will castrate a pig will depend on factors such as the age and weight of the pig, as well as their personal preference. Methods will vary in their use of anaesthetic (either local or general) and the surgical approach. Each method has its own associated benefits, risks and cost. In this way, comparing costs between veterinary practices may lead to keepers not actually comparing like for like, and instead a description of the procedure should be sought. Risks to this procedure include bleeding, bacterial infection or gut herniation, which your vet will discuss with you.

The procedure may be deemed too risky in older pigs due to the large blood vessels associated with fully grown testes. In this case, your vet may, instead, use a vaccine (Improvac (Zoetis) that acts to prevent the testicles from producing testosterone and therefore stops the pig from being fertile.

Spaying

The risk of pigs developing uterine tumours (*see* Chapter 9, 'Diseases of Pigs') is leading to enquiries from pet pig owners regarding having their female pigs spayed. This procedure is more common in the USA. Please be aware that a pig spay is certainly not standard practice for a farm vet! As described elsewhere, we cannot use the same medicines in pigs as we can in dogs and cats, and therefore even the general anaesthetic that is required is not straightforward. If you would like to investigate the procedure in the UK, we would recommend approaching a farm animal referral centre (sometimes attached to veterinary schools).

Claire's preferred method for pig castrations under 25kg is to use a general anaesthetic. This adds cost to the procedure but means that less pain is felt by the pig and allows her to hang the pig from a ladder. This means that she can visualise and sew up the inguinal canal to prevent a hernia from occurring, which would be catastrophic if it occurred after the surgery.

CHAPTER 16

PIG DEATH AND EUTHANASIA

We are incredibly lucky to be able to end a pig's pain and suffering. Despite this, the time when a pig's life comes to an end can be hugely sad for those involved. Some steps can be taken to lessen the inevitable emotional stress, which is often best achieved through prior communication with your vet and possibly other parties.

A QUESTION OF WELFARE

Welfare at the time of death is crucial to good welfare across a holding. With this in mind, it is always preferable to have a qualified and experienced professional euthanase a pig. Pigs are one of the most difficult animals to euthanase, and choosing a suitable method is dependent on several factors, such as the age and weight of the pig.

Ensuring that a professional will be available to do this requires some planning. Speak to your vet about whether they would feel comfortable to do this, and confirm that the method they would use is compatible with your wishes. If it is important, ask for some quotes to aid your decision. An alternative professional can be used, such as a licensed knackerman who will also generally be able to take away the carcass, which must be removed by an approved transporter 'as soon as is reasonably practicable'. Be sure to have the contact telephone numbers of professionals whom you have spoken to before, saved into your phone.

When choosing which professional to use for euthanasia, it is important to consider the specific situation on your holding. A suitable professional should be able to explain the process from start to finish, and alleviate your fears. Each professional will have their preferred methods, developed after much experience. Generally, as long as these methods are compatible with your wishes and are in line with science and best practice, this should be respected.

Keepers should be aware that it is illegal to bury a pig, including piglets, stillborns and afterbirth.[1] It is also illegal to burn carcasses on your own holding, without a licensed incinerator. The National Fallen Stock Company may be a useful resource.

Where possible, carcasses should be collected from the perimeter of holdings so that vehicles and their associated diseases do not need to enter your holding. This should be considered when deciding where to euthanase larger pigs; however, please also keep in mind whether a pig is in a fit state to be moved alive. Common sense and humanity must be applied here.

Legally, it is really important to differentiate between pigs that need to be euthanased as an emergency, and those that do not. A pig euthanasia should be treated as an emergency when they have 'an injury or disease associated with severe pain or suffering and do not respond to treatment'. It is then 'necessary to kill the pig as soon as possible to alleviate that pain or suffering'.[2] In effect this means immediately, and urgent euthanasia must not be delayed, meaning that it is even more important to have protocols arranged in advance. It is not acceptable for any pig that is suffering to be left to die.

In this emergency scenario only, it is legal for a keeper to euthanase a pig themselves. This can be done by any method that is 'effective and kills the pig humanely without causing any avoidable pain, distress or suffering, and the legal requirements are met'.[2] For this reason it is very much recommended that keepers are trained to have an understanding of how this can be achieved. This training can be delivered by your vet or by the Humane Slaughter Association (HSA). Training should be completed as soon as pigs are acquired, as accidents and acute disease that might require euthanasia can happen at any stage. Yearly refresher training should also be sought, as is required on all commercial pig farms in the UK, especially as smallholders will likely be doing this very infrequently and may forget important points.

Never attempt to kill a pig unless you have received suitable training.[2]

SOME METHODS USED FOR EMERGENCY EUTHANASIA

In this chapter we will go through some methods that can (or cannot) be used for emergency euthanasia. We have included this for the following reasons:

- To allow keepers to know what to expect if a qualified and experienced professional euthanases their pigs.
- To explain the reasonings behind methods and practices so that keepers understand why qualified and experienced professionals are doing certain things.
- To allow keepers to choose between methods that could be used on their holding by qualified and experienced professionals.
- To allow keepers to decide whether they would like to investigate training in euthanasia methods so they can euthanase pigs themselves.
- So that keepers can be aware of humane methods of euthanasia, because we are aware of some very inhumane methods that are used.

These explanations are not to be used as a manual for pig euthanasia. There are critical details that we have not described and therefore descriptions are unsuitable for this purpose. The law around acceptable methods of euthanasia is changeable. Therefore, up-to-date legislation must be consulted frequently by anyone euthanising pigs to ensure compliance.

EUTHANASING PIGLETS

It is not unusual to need to euthanase a piglet, for example after injury by a sow.

By Lethal injection

Euthanasia of piglets on the majority of smallholdings will be best achieved by (very promptly) taking the piglet to the vet to be euthanased by lethal injection. This is easier to achieve in piglets than larger pigs as they can be turned on their backs and their neck extended. This allows the vet to inject an overdose of an anaesthetic into a vein that runs down their neck. Any pig euthanased by lethal injection cannot then be consumed by people or animals.

By Blunt Force Trauma

Piglets under 5kg and less than four weeks of age can be euthanased by delivering a sharp, firm blow on the top of the head. If done correctly, this will cause immediate loss of consciousness and death. However, mainly due to the aesthetics of this technique but partially due to the potential for operator error, the commercial pig sector is quickly moving away from this method. The technique relies on the softer nature of the skull, due to cartilage not yet forming bone in a young piglet, so it must not be used for older or larger piglets. The technique involves holding the piglet by the back legs and swinging it very, very hard, against a solid object.

By Non-Penetrative Captive Bolt

The commercial pig sector is increasingly using an instrument called a non-penetrative captive bolt to deliver the same effect for piglets up to 10kg. This instrument uses an ammunition pellet to drive a metal disk-shaped bolt into the piglet's head. The piglet must be restrained using a device where all human body parts are kept well away from its head, and which ensures that the piglet cannot lower its head when the bolt is fired.[3]

Other Methods

Other methods that we have heard used are drowning and suffocation. These are not acceptable ways to euthanase a pig due to the suffering that they will cause before death. They are not humane and should not be practised.

LARGER PIGS

By Lethal Injection:

This method can only be used by a vet. An overdose of anaesthetic is injected to slow and stop the heart. In adult pigs this can be challenging as pigs have deep veins that lie under layers of tissue, meaning that the vet cannot access the same blood vessels as other species. Instead, they quite often inject straight into the heart or into the abdomen. To do both of these it is important that the pig is lying down, and therefore, in general, pigs must be heavily sedated (anaesthetised) first. Even when a pig is so sick that it cannot stand, we must consider the pain involved in the procedure and seriously consider sedation.

The products that vets are allowed to use for pig sedation can be very expensive, but

Poor Twiggy here is calmly being put to sleep by Sue and Stephen's farm vet, Kate, from Dyfed Farm Vets.

their use is often absolutely imperative in order to achieve a humane death by lethal injection. Furthermore the process can be long, which may lead to higher costs. Pigs take about fifteen minutes to lie down after injection with sedation. This could be longer for many reasons, such as higher stress or accidental underdosing where pigs are heavier than they look. Following this, the pig will need to be injected. Depending on which injection site the vet can access, the medication can take a variable length of time to take effect and eventually stop the heart.

Where cost could become prohibitive, it is important to speak to your vet beforehand and ascertain how they charge for pig euthanasia, understand whether they could give you a quote, and seek details on what to expect from the process.

Also bear in mind that a vet will not be able to remove the carcass for you, so you will need to arrange removal by an approved transporter who will dispose of the carcass in a legal way. Be sure to tell transporters that the pig has been euthanased by lethal injection as it must not go for consumption by any species.

Penetrative Captive Bolt Stunner

Some vets, knackermen and even some small-scale producers (with appropriate training) will use a captive bolt stunner for pig euthanasia. Penetrative captive bolt stunners do not require a firearms licence, but must only be used where training has been undertaken and competence certified.

The HSA provides a resource[4] that details everything from the positioning required for pigs, maintenance of the captive bolt, and more. This resource should be consulted, as well as appropriate training undertaken, before the procedure is completed without supervision.

A penetrative captive bolt stunner involves firing a blank cartridge that causes a 'bolt' (a metal rod) to be propelled from the device. When positioned correctly this will impact the skull and penetrate the brain, which leads to dysfunction or destruction of the brain. This leaves the pig stunned so that a final killing procedure (bleeding or 'pithing') can then be performed.

An example of a captive bolt stunner that can be used for pigs.

Penetrative captive bolt stunners must not be used for piglets under 5kg, as their skull has not fully formed to bone, which can lead to an ineffective stun. They are also best not used on pigs over about 250kg in size, as the skull can be too thick for the bolt to penetrate; therefore other means should be used.

Procedure for Use

When the procedure is performed correctly, it is very humane. However, it does look quite disturbing due to the pig making paddling movements and kicking dramatically, which can occur after it has been stunned. It is important to note that the pig has already been stunned so has no knowledge of this part of the procedure, but we still recommend that keepers do not attend. The operator will generally need someone to hold the pig on a snare, so we would advise having a fairly strong adult, who does not have a relationship with the pig, available for this.

Preparation: The operator will ensure that the captive bolt is frequently cleaned and checked, and that the other required equipment is serviced. Where possible (unless the pig is too sick to be moved), the pig will be isolated and out of sight of onlookers, in a pen suitable for restraint with a snare, and in an area where the carcass can be moved for subsequent collection.

Restraint: A captive bolt stunner is a lethal weapon at worst, or a severely life-limiting one at best. A pig must be properly restrained so that no human body parts are near the stunner, and so that the pig cannot move its head, which potentially could cause the shot to be administered in the wrong place. Unless the pig is already down and near unresponsive, this is only really possible using a snare. A pig should never be restrained (at any point, but especially for shooting with a captive bolt stunner) by holding its ear or by

The correct point of aim for a penetrative captive bolt stunner in a pig is at the intersection of these two lines.

putting its head between the knees. This is an incredibly dangerous practice.

Delivering the stun: Once restrained, the pig will be shot in the centre of its forehead. From this moment, as long as the procedure is completed correctly and the pig is killed without delay, the pig is completely unconscious.

Procedure to Kill the Pig

Legally, the firing of the device is a stunning procedure, not a killing procedure. This means that, following the shot, a procedure to actually kill the pig is required. This is required by law and, more importantly, prevents the pig from regaining consciousness.

This can be done by a process called pithing, or bleeding the pig. Pithing is less messy so is generally preferred for emergency euthanasia. The procedure involves putting a rod into the hole created by the stunner and moving it back and forth to ensure that the brain stem has been destroyed. This might sound an appalling thing to do, however we can firmly tell you that nothing is more horrible than a pig regaining consciousness after it has been shot, and therefore it is imperative to do this every time.

Straight after firing the pig will drop to the floor, completely rigid and very slightly shaking. Here, the operator has five to ten seconds to pith before the pig starts to exhibit (often very dramatic) kicking and paddling movements, and moves across the ground.

Once the pig has been pithed and kicking and paddling ensues, the operator will monitor the pig to check that the procedure has been effective. If they are in any doubt, they will reshoot (slightly away from the first shot) and start the process again.

FIREARMS AND SHOTGUNS

In order to own a gun, a firearms certificate or shotgun licence must be obtained from the police. These can take considerable time to arrive and therefore alternative euthanasia methods must be available until then. Once a certificate or licence has been granted, the operator must be trained and certified to use the weapon for pig euthanasia before using it for this purpose. A shotgun (12 or 20 bore) is far preferable to a rifle because the risk to the operator is lower.

The procedure with a rifle or shotgun is different from the use of a penetrative captive bolt gun in the following ways:

- The muzzle must not be placed next to the skull, but instead should be 5 to 20cm away.
- Even more care must be taken to prepare the area that allows correct positioning and human safety. This includes considering ricochets and misfires. Pigs should be shot on soft ground, have a safe backdrop, and not be shot in an enclosed space.
- Pithing after firing is not required.

That being said, the procedure is very similar to a captive bolt stunner in terms of the following:

- It is the same point of aim on the pig's forehead.
- There is the same rigid phase followed by involuntary movement in the pig after shooting.
- There are the same signs to look out for in terms of whether the procedure is successful or not.

CAN AN INJURED PIG STILL BE ACCEPTED BY AN ABATTOIR?

There is a specific set of circumstances when a pig is allowed to be killed on farm and can still be accepted by an abattoir, or is allowed to be sent to an abattoir if it is sick or injured; however, these cases are few and far between.[5, 6]

The keeper will need to fill out additional food-chain information detailing written declarations. If the pig is alive it must still be legally fit for transport and travel in a separate compartment in a deep bed to the nearest abattoir. If the pig is to be killed on the holding, a vet must attend to complete a certificate that will, in part, confirm that the injury was very acute, that it is not within a medicine withdrawal period, and that there are no pre-mortem concerns in terms of the meat. Then the slaughter must be carried out by a licensed slaughterman - but this must not cause a delay in the humane euthanasia of the pig.[7] Delay must also not have been caused in getting the pig treated for injury or disease, and treatment must not have been withheld where it was necessary. In addition to this, many abattoirs will not accept pigs killed on the holding.

Therefore it is generally inappropriate to send an injured pig to an abattoir, either dead or alive, and pig keepers are urged not to pressure their vet to sign certificates for this. Inappropriate certification could lead to prosecution due to food safety concerns.

NON-EMERGENCY EUTHANASIA

Non-emergency cases can only be euthanased by a keeper after achieving further certificates of competency, which can vary for different authorities. Home slaughter has separate requirements that are outside the scope of this book, but readers can consult the HSA for more information.[8]

CONCLUSION

This book has been a labour of love for all the contributors, and we hope you have found it instructive, entertaining and perhaps even enjoyable. We thank all those involved, especially our wonderful writers of breed summaries, those who have gifted us fantastic photos and those who have taught us the hints and tips over the years that we have now been able to pass to you. Without them, this project would not be what it is.

We also thank those in the sector who have provided amazing resources that we have extensively drawn upon as references. Through this, we have been able to consider how tried and tested strategies could be adapted for the smallholding audience, and we have strived to ensure that we provide readers with correct and evidenced material at the time of writing.

As vets and farmers our priority is the welfare and quality of life of the animals committed to our care, and with that in mind we set out to create a practical textbook that could be read in its entirety by the novice or used as a convenient reference by the experienced pig keeper. We hope our enthusiasm shines through these pages.

Enjoy your pigs!

Claire, James, Pete, Stephen, and Sue.

GLOSSARY

There are many throw-away terms that you will hear used. This glossary should give you a guide.

abortion Birth of dead piglets before 109 days into pregnancy.
acute Sudden onset of disease.
AHDB Pork Agriculture and Horticulture Development Board Pork is the levy board for pig producers. They take a financial cut of every pig slaughtered in the UK, which they use for research to ensure that we are using best pig practices and to share best practice with producers, for example by running events and creating resources. They are responsible for key initiatives such as the electronic medicines book, the module on moving and handling, and the significant diseases charter.
African swine fever A highly contagious and notifiable disease of pigs with commonly a 100 per cent mortality rate. It has not yet reached the UK, but is spreading at a rate of approximately 100km a year and has already reached Belgium.
agalactia A cessation of milk production in one or more teats.
arc Outdoor accommodation for pigs. They come in a variety of designs and sizes.
baconer A pig reared to produce bacon and slaughtered at around 80-100kg or eight to ten months of age.
bagging up Increase in the size of the teats prior to milk production.
barrow or hog Castrated male.
biosecurity The practice of preventing the spread of disease from one holding to another or within a holding.
boar An uncastrated male pig of more than six months of age.
BPA British Pig Association.
cake or rolls Pig food made in large cylindrical shapes which are ideal for feeding from the ground.
chronic Slower onset of disease.
Code of Practice (for the welfare of pigs) A document that keepers are required to have a copy of, and be familiar with, by law. It can be found at https://assets.publishing.service.gov.uk/government/uploads/system/uploads/attachment_data/file/908108/code-practice-welfare-pigs.pdf
colostrum A very rich milk containing antibodies that is produced by the sow at farrowing and is essential for newborn piglets.
concentrate (noun) Grains or processed, pelleted pig feed, also known as hard feed.
creep area Subdivision of the farrowing area that piglets can access but the sow can't. It generally has a heat lamp and is a safe, warm area for the piglets.
creep feed Highly palatable and easily digestible feed that may be offered to piglets while suckling, normally at around ten to fourteen days of age.
cull To remove a pig from the herd, not necessarily by killing.
cutter A pig reared to produce larger joints of meat, between a porker and baconer in size.
dead weight The weight of a pig after slaughter once the pluck etc has been removed.
dry sow A sow that is not lactating.
drying off The process of stopping a sow's lactation after weaning.
endemic A disease persistently present in a region or population.
erysipelas Common bacterial disease of pigs, often with characteristic diamond marks on the skin.
farrow To give birth.
farrowing crate A means of confining the sow for a period after farrowing in order to protect the piglets. Most often used in intensive systems.
farrowing rails Rails fitted within a farrowing arc to help prevent piglets being lain on by the sow. They fulfil a similar role to a creep area.
fender A barrier used to stop piglets straying too far from their arc. Used in outdoor farrowing.
finisher Meat pig in the last part of its life.
forage (noun) Leafy plant matter as feed, such as grass, hay, haylage or silage.
gestation The length of a pregnancy. Generally this is three months, three weeks and three days (115 days).

gilt Young female pig who has not yet had a litter.
grower Young meat pig older than a weaner but still in the earlier part of its life.
hard feed Grains or processed, pelleted pig feed, also known as concentrates.
hog or barrow Castrated male.
hogging Female pig in season / on heat.
HSA humane slaughter association.
in pig A pregnant sow.
lice Large and easily seen ectoparasites.
litter A group of sibling piglets born from the same pregnancy.
live weight The weight of a live pig.
mange Skin disease caused by parasitic mites.
mastitis Inflammation of a teat, normally causing it to become non-functional.
meat pigs Pigs destined to be reared only for meat, not for breeding.
metritis Inflammation of the uterus.
nesting The natural instinctive behaviour of a sow just prior to farrowing where she will gather bedding materials to make a nest.
notching A form of identification used in pedigree pigs, where small notches are clipped into the ear in a pattern which corresponds with the pig's tag number.
notifiable disease A disease that, by statutory requirements, must be reported to the public health or veterinary authorities when the diagnosis is made because of its importance to human or animal health.
parasites Organisms living on or inside other creatures and dependent upon them: for example intestinal worms, mange mites or pig lice.
parity The number of litters a sow has carried (including the current pregnancy); for example a 'second parity sow' has weaned her second litter but not yet given birth to her third.
pathogen A microorganism (bacteria, virus, parasite etc) that can cause disease.
pencils Pig food made in a fine cylindrical shape making it easy for piglets and weaners to eat.
pig tipping Tickling a pig behind the front leg or in front of the hind leg until it falls over. This is very enjoyable for all participants.
piglet A young pig still suckling from the sow.
pluck The lungs, heart, liver etc - edible organs that can be requested from the abattoir.
porker A pig reared to produce pork. This is normally around 60kg, and reached between four and six months of age depending on the breed.
rolls or cake Pig food made in large cylindrical shapes that are ideal for feeding from the ground.
root (verb) To dig up the ground with the snout, looking for food.
sausage Apart from the obvious, it also describes the raised mound below a sow's belly on which the teats sit. It develops towards the end of the gestation period.
slap mark A type of tattoo used for marking pigs with their herd number prior to slaughter.
slapshot A tube that fits between a syringe and a hypodermic to make injecting pigs simpler when they move.
slaughter tag A small metal tag put on to the edge of an ear, with the owner's herd number inscribed on it. These are used for slaughter as an alternative to plastic tags, which might not necessarily survive the scalding process.
slurry Faeces and urine mixed, or not, with some litter material and some water to give a liquid manure with a dry matter content up to about 10 per cent that flows under gravity and can be pumped.[1]
snare A sliding loop of rope or wire used for restraining a pig around its upper jaw.
sow Female pig that has had a litter.
standing heat When a hogging sow stands still to pressure on her back, meaning she is ready to be served. A traditional method of checking for standing heat was to try riding the sow - if she tolerated being ridden she was ready for the boar. This is not a method we recommend.
tagging The standard method of identifying pigs - a plastic tag through the ear with the owner's herd number and, in the case of pedigree pigs, the pig's individual number as well (*see* **slaughter tag**).
tattooing An alternative form of identification to notching for pedigree pigs, where the pig's number is tattooed into an ear. Can only be used on pigs with a pink ear.
to market Sending a pig to an abattoir for slaughter.
trichinella A parasite of pigs that can be harmful to humans if ingested. It is controlled by the meat of pigs that could be infected (outdoor pigs) being tested at slaughter, which is a legal requirement in the UK.
tusks Large, sideways-pointing teeth on the lower jaw of mature boars. They can be very sharp and should trimmed for the safety of both handlers and sows.
underline The number and placement of a pig's teats on both males and females. It is a very important factor when choosing breeding animals.
wallow A muddy pool that pigs can use to cool themselves in summer.
weaner Young pig that has been weaned from the sow.
weaning The act of removing piglets from the sow in preparation for independence.

REFERENCES AND RESOURCES

CHAPTER 1

[1]Defra (2020) Code of practice for the welfare of pigs. https://assets.publishing.service.gov.uk/government/uploads/system/uploads/attachment_data/file/908108/code-practice-welfare-pigs.pdf. Accessed 28 July 2021.

[2]FAWC (2007) FAWC report on stockmanship and farm animal welfare. https://www.gov.uk/government/publications/fawc-report-on-stockmanship-and-farm-animal-welfare. Accessed 01 April 2023.

[3]AHDB (2020) Evidence report: comparing the potential implications of widespread use of different farrowing systems in the British pig sector. https://projectblue.blob.core.windows.net/media/Default/Market%20Intelligence/COP/AHDB%20Alternative%20Farrowing%20Report.pdf. Accessed 21 August 2022.

[4]Rare Breeds Survival Trust (2022) Pigs watchlist. https://www.rbst.org.uk/pages/category/pigs-watchlist?Take=11. Accessed 28 Nov 2022.

[5]European Commission (2024) Organic production and products. https://agriculture.ec.europa.eu/farming/organic-farming/organic-production-and-products_en. Accessed 01 April 2023.

[6]Soil Association (2022) Soil association standards farming and growing. https://www.soilassociation.org/media/23372/sa-gb-farming-_growing-standards.pdf. Accessed 05 Nov 2022.

[7]Red Tractor (2022) Pigs Standards. https://redtractorassurance.org.uk/wp-content/uploads/2022/08/Pigs-V5.1-Standards-FINAL.pdf. Accessed 20 August 2022.

[8]RSPCA (2016 plus supplementary material in 2021 and 2022). RSPCA welfare standards for pigs. https://science.rspca.org.uk/sciencegroup/farmanimals/standards/pigs. Accessed 20 August 2022.

[9]AHDB (2022) Pork provenance pig production methods. https://www.porkprovenance.co.uk/productionmethods.asp#%20. Accessed 20 August 2022.

[10]The Veterinary Surgeons Act 1966. https://www.legislation.gov.uk/ukpga/1966/36. Accessed 10 Sept 2022.

CHAPTER 2

[1]Defra (2017) Guidance: register land you use to keep livestock. https://www.gov.uk/guidance/register-land-you-use-to-keep-livestock. Accessed: 28 July 2021.

[2]The Pigs (Records, Identification and Movement) Order 2011. https://www.legislation.gov.uk/uksi/2011/2154/contents. Accessed 28 July 2021.

3 The Welfare of Animals (Transport) (England) Order 2006. https://www.legislation.gov.uk/uksi/2006/3260/contents. Accessed 28 July 2021.

[4]Defra (2012) Guidance: Pig keepers: report and record movements to or from your holding. https://www.gov.uk/guidance/pig-keepers-report-and-record-movements-to-or-from-your-holding. Accessed: 28 July 2021.

[5]Defra (2014) Guidance: supplying and using animal by-products as farm animal feed. https://www.gov.uk/guidance/supplying-and-using-animal-by-products-as-farm-animal-feed. Accessed 28 July 2021.

[6]Defra and APHA (2014) Guidance: African swine fever: how to spot and report the disease. Available at: https://www.gov.uk/guidance/african-swine-fever. Accessed 16 September 2021.

[7]The Welfare of Farmed Animals (England) Regulations 2007. https://www.legislation.gov.uk/uksi/2007/2078/regulation/6/made. Accessed 28 July 2021.

[8]The Veterinary Medicines Regulations 2013. https://www.legislation.gov.uk/uksi/2013/2033/regulation/19/made. Accessed 01 April 2023.

[9]AHDB Pork (2021) The electronic medicine book for pigs. https://emb-pigs.ahdb.org.uk/. Accessed 04 August 2021.

[10]Defra (2020) Code of practice for the welfare of pigs. https://assets.publishing.service.gov.uk/government/uploads/system/uploads/attachment_data/file/908108/code-practice-welfare-pigs.pdf. Accessed 28 July 2021.

[11]Animal Health Act 1981. Section 15. https://www.legislation.gov.uk/ukpga/1981/22/section/15. Accessed 10 August 2021.

[12]Defra (2012b) Guidance: fallen stock and safe disposal of dead animals. https://www.gov.uk/guidance/fallen-stock. Accessed 28 July 2021.

[13]Pig Veterinary Society (2013) The casualty pig. https://www.pigvetsoc.org.uk/files/document/192/Casualty%20Pig%20-%20April%202013-1.pdf. Accessed 28 Jul 2021.

[14]Natural England and Defra (2014) EIA (Agriculture) regulations: apply to make changes to rural land. https://www.gov.uk/guidance/eia-agriculture-regulations-apply-to-make-changes-to-rural-land. Accessed 31 July 2022.

[15]NFU (2013) Pig industry good practice checklist reducing odours from pig production through the application of best available techniques. https://www.nfuonline.com/archive?treeid=17335. Accessed 20 August 2022.

[16]Food Standards Agency (2020) Animal feed legislation. https://www.food.gov.uk/business-guidance/animal-feed-legislation. Accessed 31 July 2022.

CHAPTER 3

[1]Defra (2020) Code of practice for the welfare of pigs. https://assets.publishing.service.gov.uk/government/uploads/system/uploads/attachment_data/file/908108/code-practice-welfare-pigs.pdf. Accessed 28 July 2021.

[2]AHDB (2022) Significant diseases charter. https://ahdb.org.uk/significant-diseases-charter. Accessed 19 August 2022.

[3]AHDB (2022) Pork order form. https://ahdb.org.uk/pork-order-form. Accessed 19 August 2022.

[4]Red Tractor (2021) Pigs Standards. https://redtractorassurance.org.uk/wp-content/uploads/2021/08/RTStandardsV5_PIGS_SINGLES.pdf. Accessed 19 August 2022.

[5]Defra (2022) Disinfectants Approved for use in England, Scotland and Wales. http://disinfectants.defra.gov.uk/DisinfectantsExternal/Default.aspx?Module=ApprovalsList_SI. Accessed 19 August 2022.

CHAPTER 4

[1]Farm Animal Welfare Council (2007). FAWC Report on stockmanship and farm animal welfare. https://assets.publishing.service.gov.uk/government/uploads/system/uploads/attachment_data/file/325176/FAWC_report_on_stockmanship_and_farm_animal_welfare.pdf. Accessed 25 July 2022.

[2]The Welfare of Farmed Animals (England) Regulations 2007. https://www.legislation.gov.uk/uksi/2007/2078/regulation/6/made. Accessed 28 July 2021.

[3]AHDB (2022) How to manage the farrowing arc or hut. https://ahdb.org.uk/knowledge-library/how-to-manage-the-farrowing-arc-or-hut. Accessed 31 July 2022.

[4]Blair R (2018) Nutrition and feeding of organic pigs 2nd edition. CABI.

[5]SRUC (2019) Outdoor Pig Husbandry: A stockperson's guide to farrowing. https://www.sruc.ac.uk/media/5jliy2wd/outdoor-farrowing-manual.pdf. Accessed 28 Nov 2022.

[6]Defra (2020) Code of practice for the welfare of pigs. https://assets.publishing.service.gov.uk/government/uploads/system/uploads/attachment_data/file/908108/code-practice-welfare-pigs.pdf. Accessed 28 July 2021.

[7]AHDB (2022) Weather, season and pig breeding. https://ahdb.org.uk/knowledge-library/weather-seasons-and-pig-breeding. Accessed 02 August 2022.

Common disease conditions diagnosed in smallholder and pet pigs. http://apha.defra.gov.uk/documents/surveillance/diseases/disease-smallholder-pet-pig.pdf. Accessed 27 August 2022.

[9]AHDB (2022) Indoor farrowing body condition in sows. https://ahdb.org.uk/knowledge-library/indoor-farrowing-body-condition-in-sows. Accessed 21 August 2022.

[10]AHDB (2022) Temperature requirements for pigs. https://ahdb.org.uk/knowledge-library/temperature-requirements-of-pigs. Accessed 21 August 2022.

[11]RSPCA (2016) RSPCA welfare standards for pigs. https://science.rspca.org.uk/sciencegroup/farmanimals/standards/pigs. Accessed 20 August 2022.

[12] Soil Association (2022) Soil association standards farming and growing. https://www.soilassociation.org/media/23372/sa-gb-farming-_growing-standards.pdf. Accessed 05 Nov 2022.

CHAPTER 5

[1]Sardesai, V. M. (1992) Nutritional role of polyunsaturated fatty acids. The Journal of Nutritional Biochemistry. 3(4), 154-166.

[2]Blair R (2018) Nutrition and feeding of organic pigs 2nd edition. CABI.

[3]Barb CR, Hausman CJ, Houseknecht KL (2001) Biology of leptin in the pig. Domestic Animal Endocrinology. 21(297-217).

[4]Defra (2014) Guidance: supplying and using animal by-products as farm animal feed. https://www.gov.uk/guidance/supplying-and-using-animal-by-products-as-farm-animal-feed. Accessed 28 July 2021.

[5]UfacUK (unknown) Sustainable proten source offers viable alternative to soybean meal and fishmeal in pig diets. https://www.ufacuk.com/news/sustainable-protein-source-offers-viable-alternative-to-soybean-meal-and-fishmeal-in-pig-diets/. Accessed 02 August 2022.

[6]Edwards S (2002) Feeding organic pigs: A handbook of raw materials and recommendations for feeding practice. https://orgprints.org/id/eprint/38590/6/38590_Tool_90_Ok-Net-Ecofeed_feeding-organic-pigs-a-handbook.pdf. Accessed 01 August 2022.

[7]Zimmerman JJ, Locke AK, Ramirez A, Schwartz KJ, Stevenson GW, Zhang J. (2019) Diseases of Swine Eleventh Edition. Wiley Blackwell.

[8]AHDB (2022g) Removing zinc oxide from pig diets. https://ahdb.org.uk/knowledge-library/removing-zinc-oxide-from-pig-diets. Accessed 02 August 2022.

[9]De Leeuw JA., Bolhuis JE., Bosch G, Gerrits, WJJ (2008) Effects of dietary fibre on behaviour and satiety in pigs: symposium on 'Behavioural nutrition and energy balance in the young'. Proceedings of the Nutrition Society, 67(4), pp.334-342.

[10]AHDB (2022) Mycotoxins in pig feed. https://ahdb.org.uk/knowledge-library/mycotoxins-in-pig-feed. Accessed 02 August 2022.

[11]AHDB (2022) Feeding gestating (dry) sows. https://ahdb.org.uk/knowledge-library/feeding-gestating-dry-sows. Accessed 02 August 2022.

[12]Allen & Page (2018) Smallholder range pigs: Palatable and nutritious feeds for pigs. . https://www.smallholderfeed.co.uk/products/pigs/. Accessed 02 August 2022.

[13]Allen, A. D., Lasley, J. F., & Tribble, L. F. (1959). Milk production and related performance factors in sows. University of Missouri, College of Agriculture, Agricultural Experiment Station.

[14]AHDB (2020) Evidence report: Comparing the potential implications of widespread use of different farrowing systems in the British pig sector. https://projectblue.blob.core.windows.net/media/Default/Market%20Intelligence/COP/AHDB%20Alternative%20Farrowing%20Report.pdf. Accessed 21 August 2022.

[15]AHDB (2022) Feeding lactating sows. https://ahdb.org.uk/knowledge-library/feeding-lactating-sows. Accessed 02 August 2022.

[16]Farmers Weekly (2021) Video: Norfolk pig farmers showcase creep-feeder inventions. https://www.fwi.co.uk/livestock/pigs/video-norfolk-pig-farmers-showcase-creep-feeder-inventions. Accessed 02 August 2022.

[17]AHDB (2022) Body condition scoring sows. https://ahdb.org.uk/knowledge-library/body-condition-scoring-sows. Accessed 01 August 2022.

[18]SRUC (2019). Outdoor Pig Husbandry: A stockperson's guide to farrowing. https://www.sruc.ac.uk/media/5jliy2wd/outdoor-farrowing-manual.pdf. Accessed 28 Nov 2022.

[19]EUPiG (2020) EU PiG Innovation Group: Technical Report - Welfare. https://ec.europa.eu/research/participants/documents/downloadPublic?documentIds=080166e5d2d55404&appId=PPGMS. Accessed 21 August 2022.

[20]AHDB (2022) Gilt service to farrowing. https://ahdb.org.uk/knowledge-library/gilt-service-to-farrowing. Accessed 02 August 2022.

[21]Cotswold Seeds (2022) Pig rooting mix. https://www.cotswoldseeds.com/products/2506/pig-rooting-mix-short-term-1-2-years. Accessed 23 October 2022.

[22]Soil Association (2022) The impacts of nitrogen pollution. https://www.soilassociation.org/causes-campaigns/fixing-nitrogen-the-challenge-for-climate-nature-and-health/the-impacts-of-nitrogen-pollution/. Accessed 21 August 2022.

[23]Pollock C. (2019) Body condition scoring the miniature pig. Available at https://lafeber.com/vet/body-condition-scoring-the-miniature-pig/. Accessed 10 Sept 2022.

CHAPTER 6

[1]Defra (2020) Code of practice for the welfare of pigs. https://assets.publishing.service.gov.uk/government/uploads/system/uploads/attachment_data/file/908108/code-practice-welfare-pigs.pdf. Accessed 28 July 2021.

[2] Grandin (2012) https://www.templegrandin.com/. Accessed 05 November 2022.

[3] Schenkel R (1947) Expression studies on wolves: Captivity observations.

[4]AHDB (2021) MODULE 1: moving & handling pigs. https://ahdbpork.vbms.co.uk/store. Accessed 13 August 2022.

[5]AHDB (2020) Work instruction: handling and restraining pigs. https://projectblue.blob.core.windows.net/media/Default/Pork/Documents/Work%20instructions/WI%20-%20Restraining%20pigs.docx. Accessed 05 August 2021.

CHAPTER 7

[1]The Welfare of Animals (Transport) (England) Order 2006. https://www.legislation.gov.uk/uksi/2006/3260/contents. Accessed 28 July 2021.

[2]The Pigs (Records, Identification and Movement) Order 2011. https://www.legislation.gov.uk/uksi/2011/2154/contents. Accessed 28 July 2021.

[3]Defra and APHA (2022) Guidance: animal welfare, advice and guidance on protecting animal welfare on farms, in transport, at markets and at slaughter. https://www.gov.uk/guidance/animal-welfare. Accessed 14 August 2022.

[4]European Commission (2017) Consortium of the animal transport guides project: guide to practices for the transport of pigs. http://animaltransportguides.eu/wp-content/uploads/2016/05/D3-Pigs-Revised-Final.pdf. Accessed 14 August 2022.

[5]Defra (2020) Code of practice for the welfare of pigs. https://assets.publishing.service.gov.uk/government/uploads/system/uploads/attachment_data/file/908108/code-practice-welfare-pigs.pdf. Accessed 28 July 2021.

[6]European Parliament (2005) Regulation (EC) No 1/2005 on the protection of animals during transport and related operations. https://www.europarl.europa.eu/RegData/etudes/STUD/2018/621853/EPRS_STU(2018)621853_EN.pdf. Accessed 13 August 2022.

[7]AHDB (2021) MODULE 1: Moving & handling pigs. https://ahdbpork.vbms.co.uk/store. Accessed 13 August 2022.

[8]UECBV Eurogroup for animals, FVE, INAPORC, COPA COGNECA, COOPERL ELT, ANIMALS ANGELS, IRU (2016) Practical guidelines to assess fitness for transport of pigs. http://eurocarne.com/daal/a1/informes/a2/Practicalguidelinestoassessfitnessoftransportofpigs.pdf. Accessed 14 August 2022.

CHAPTER 8

[1]Animal Welfare Act (2006) https://www.legislation.gov.uk/ukpga/2006/45/contents. Accessed 10 Sept 2022.

[2]Grisha Stewart (2022) Cooperative care streaming video series. https://store.grishastewart.com/products/cooperative-care. Accessed 07 Sept 2022.

[3]FAWC (2007) FAWC report on stockmanship and farm animal welfare. https://assets.publishing.service.gov.uk/government/uploads/system/uploads/attachment_data/file/325176/FAWC_report_on_stockmanship_and_farm_animal_welfare.pdf. Accessed 25 July 2022.

[4]Navarro E, Mainau E, Manteca X (2020) Development of a facial expression scale using farrowing as a model of pain in sows. Animals. 14;10(11):2113.

CHAPTER 9

[1]Veterinary Surgeons Act 1966. https://www.legislation.gov.uk/ukpga/1966/36. Accessed 29 August 2022.

[2]Taylor DJ (2013) Pig diseases: Ninth Edition. 31 North Birbiston Road, Lennoxtown.

[3]The Diseases of Swine Regulations 2014. https://www.legislation.gov.uk/uksi/2014/1894/contents/made. Accessed 27 August 2022.

[4]Animal Health Act 1981. Section 15. https://www.legislation.gov.uk/ukpga/1981/22/section/15. Accessed 10 August 2021.

[5]The Pig Site (2021) ASF ravages Northern China's pig herd. https://www.thepigsite.com/news/2021/04/asf-ravages-northern-chinas-pig-herd. Accessed 27 August 2022.

[6]Defra and APHA (2014) African swine fever: how to spot and report the disease. https://www.gov.uk/guidance/african-swine-fever. Accessed 27 August 2022.

[7]Hodal K (2019) Meat infected by African swine fever found in UK for first time. https://www.theguardian.com/environment/2019/jul/11/meat-infected-by-african-swine-fever-found-in-uk-for-first-time. Accessed 27 August 2022.

[8]The Pirbright Institute and APHA (2021) African swine fever virus. Available at: https://www.pirbright.ac.uk/asfv. Accessed 16 September 2021.

[9]Defra and APHA (2014) Classical swine fever: how to spot and report the disease. https://www.gov.uk/guidance/classical-swine-fever. Accessed 27 August 2022.

[10]World Organisation for Animal Health (2022) Classical Swine Fever. https://www.woah.org/en/disease/classical-swine-fever/. Accessed 27 August 2022.

[11]Wilson J (2002) Foot and mouth farmer banned for 15 years. https://www.theguardian.com/uk/2002/jun/29/footandmouth.jamiewilson#:~:text=The%20pig%20farmer%20accused%20of,farm%20animals%20for%2015%20years. Accessed 27 August 2022.

[12]Defra and APHA (2014) Foot and mouth disease: how to spot and report it. https://www.gov.uk/guidance/foot-and-mouth-disease. Accessed 27 August 2022.

[13]Defra and APHA (2014) Swine vescicular disease: how to spot and report it. https://www.gov.uk/guidance/swine-vesicular-disease. Accessed 27 August 2022.

[14]Defra (2019) Aujeszky's disease in domestic swine in France. https://www.gov.uk/government/publications/aujeszkys-disease-in-domestic-swine-in-france. Accessed 27 August 2022.

[15]Defra and APHA (2014) Aujeszky's disease: how to spot and report it. https://www.gov.uk/guidance/aujeszkys-disease. Accessed 27 August 2022.

[16]Defra and APHA (2014) Teschen disease: how to spot it and report it. https://www.gov.uk/guidance/teschen-disease. Accessed 27 August 2022.

[17]AHDB (2022) Porcine epidemic diarrhoea virus (PEDv). https://ahdb.org.uk/knowledge-library/porcine-epidemic-diarrhoea-virus-pedv. Accessed 27 August 2022.

[18]AHDB (2022) Significant Diseases Charter. https://ahdb.org.uk/significant-diseases-charter. Accessed 27 August 2022.

[19]AHDB (2022) How to prevent bovine tuberculosis in pigs. https://ahdb.org.uk/knowledge-library/bovine-tuberculosis-in-pigs. Accessed 27 August 2022.

[20]APHA (2018) Common disease conditions diagnosed in smallholder and pet pigs. http://apha.defra.gov.uk/documents/surveillance/diseases/disease-smallholder-pet-pig.pdf. Accessed 27 August 2022.

[21]EUPiG (2020) EU PiG Innovation Group: Technical Report - Welfare. https://ec.europa.eu/research/participants/documents/downloadPublic?documentIds=080166e5d2d55404&appId=PPGMS. Accessed 21 August 2022.

[22]APHA (2018) Bracken poisoning in pigs. http://apha.defra.gov.uk/documents/surveillance/diseases/bracken-poisoning-pigs.pdf. Accessed 29 August 2022.

[23]APHA (2016) Klebsiella septicaemia - Information for pig keepers and vets. http://apha.defra.gov.uk/documents/surveillance/diseases/klebsiella-vets.pdf. Accessed 27 August 2022.

[24]AHDB (2022) Distribution and biology of black nightshade in the UK. https://ahdb.org.uk/knowledge-library/distribution-and-biology-of-black-nightshade-in-the-uk. Accessed 29 August 2022.

[25]RUMA (2010) Anthelmintics in Pigs. https://www.ruma.org.uk/wp-content/uploads/2014/09/RUMA-Pig-Anthelmintics_pigs_Long_2010.pdf. Accessed 28 August 2022.

[26]APHA (2018) Porcine reproductive and respiratory syndrome diagnoses in GB: 2016-17. http://apha.defra.gov.uk/documents/surveillance/diseases/pub-prrs.pdf. Accessed 28 August 2022.

[27]Zimmerman JJ, Locke AK, Ramirez A, Schwartz KJ, Stevenson GW, Zhang J. (2019) Diseases of Swine Eleventh Edition. Wiley Blackwell.

[28]NADIS (2010) Milking Problems in Sows and Gilts. https://www.nadis.org.uk/disease-a-z/pigs/milking-problems-in-sows-and-gilts/. Accessed 28 August 2022.

[29]Chigerwe M, Shiraki R, Olstad EC, Angelos JA, Ruby AL, Westropp JL (2013) Mineral composition of urinary calculi from potbellied pigs with urolithiasis: 50 cases (1982-2012). Journal of the American Veterinary Medical Association. 1;243(3):389-93.

[30]Maes DG, Vrielinck J, Millet S, Janssens GP, Deprez P (2004) Urolithiasis in finishing pigs. The veterinary journal. 1;168(3):317-22.

[31]Jagdale A, Iwase H, Klein EC, Cooper DK (2019) Incidence of neoplasia in pigs and its relevance to

clinical organ xenotransplantation. Comparative Medicine. 1;69(2):86-94.
[32]Haddad JL, Habecker PL (2012) Hepatocellular carcinomas in Vietnamese pot-bellied pigs (Sus scrofa). Journal of Veterinary Diagnostic Investigation. 24(6):1047-51.
[33]Horak V, Palanova A, Cizkova J, Miltrova V, Vodicka P, Kupcova Skalnikova H (2019) Melanoma-bearing libechov minipig (MeLiM): the unique swine model of hereditary metastatic melanoma. Genes. 10(11):915.
[34]Wood P, Hall JL, McMillan M, Constantino-Casas F, Hughes K (2020) Presence of cystic endometrial hyperplasia and uterine tumours in older pet pigs in the UK. Veterinary Record Case Reports. 8(1):e000924.

CHAPTER 10

[1]Veterinary Medicines Directorate. Product Information Database. https://www.vmd.defra.gov.uk/productinformationdatabase/. Accessed 25 August 2022.
[2]The Veterinary Surgery (Exemptions) Order 1962. https://www.legislation.gov.uk/uksi/1962/2557/made. Accessed 23 October 2022.
[3]PHWC (2018) Practical Guide to Responsible Use of Antibiotics on Pig Farms. phwc-good-practice-guide-sep18-final-v2.pdf (windows.net). Accessed 25 August 2022.
[4]Thepigsite.com (2002) Weighing a Pig Without a Scale. https://www.thepigsite.com/articles/weighing-a-pig-without-a-scale. Accessed 05 August 2021.
[5]MS Schippers (2021) Nylon Hypodermic Syringe, 10 ml. https://www.msschippers.co.uk/nylon-hypodermic-syringe-luer-lock-M1406401.html. Accessed 05 August 2021.
[6]York Vet Supplies (2021) Slap-Shot Flexible Vaccinator. Available at: https://www.yorkvetsupplies.co.uk/product/slap-shot-flexible-vaccinator/. Accessed 05 August 2021.
[7]Too HL, Sheikh Omar AR, Vidyadaran MK, and Toh I (1995) Iatrogenic sciatic nerve monoparalysis in pigs.
[8]Red Tractor Assured Food Standards (2021) https://redtractorassurance.org.uk/wp-content/uploads/2021/09/Broken-Needle-Policy-.docx. Accessed 14 August 2022.

CHAPTER 11

[1]Blair R (2018) Nutrition and feeding of organic pigs 2nd edition. CABI.
[2]NOAH (2022) Gletvax 6. https://www.noahcompendium.co.uk/?id=-457477. Accessed 21 August 2022.
[3]APHA (2018) Common disease conditions diagnosed in smallholder and pet pigs. http://apha.defra.gov.uk/documents/surveillance/diseases/disease-smallholder-pet-pig.pdf. Accessed 23 August 2022.
[4]NADIS (2022) Clostridium noyvi infection. https://www.nadis.org.uk/disease-a-z/pigs/clostridium-novyi-infection/. Accessed 23 August 2022.

CHAPTER 12

[1]Macrelli M, Williamson S, Mitchell S, Pearson R, Andrews L, Morrison AA, Nevel M, Smith R, Bartley DJ (2019) First detection of ivermectin resistance in Oesophagostomum dentatum in pigs. Veterinary parasitology. 1;270:1-6.
[2]RUMA (2010) Anthelmintics in Pigs. https://www.ruma.org.uk/wp-content/uploads/2014/09/RUMA-Pig-Anthelmintics_pigs_Long_2010.pdf. Accessed 28th August 2022.

CHAPTER 13

[1]AHDB (2022) Gilt service to farrowing. https://ahdb.org.uk/knowledge-library/gilt-service-to-farrowing. Accessed 02 August 2022.
[2]British Pig Association. Traditional breeds AI scheme - Guide to artificial insemination. http://deerpark-pigs.com/wp-content/uploads/2013/04/AI-instructions.pdf. Accessed 29 August 2022.
[3]Le Dividich J, Rooke JA, Herpin P (2005) Nutritional and immunological importance of colostrum for the new-born pig. The Journal of Agricultural Science. 143(6):469-85.
[4]Blair R (2018) Nutrition and feeding of organic pigs 2nd edition. CABI.
[5]Šterzl J, Rejnek J, Trávníček J (1966) Impermeability of pig placenta for antibodies. Folia microbiologica. 11:7-10.
[6]Allen, A. D., Lasley, J. F., & Tribble, L. F (1959) Milk production and related performance factors in sows. University of Missouri, College of Agriculture, Agricultural Experiment Station.

[7]Muns R, Nuntapaitoon M, and Tummaruk P (2017) Effect of oral supplementation with different energy boosters in newborn piglets on pre-weaning mortality, growth and serological levels of IGF-I and IgG. Journal of animal science. 95(1):353-360.

[8]PIC (2018) Colostrum Management Makes the Difference: For More Piglets and For Better Piglets. https://cn.pic.com/wp-content/uploads/sites/12/2018/12/8.28.18-Colostrum-Management.pdf. Accessed 31 August 2022.

CHAPTER 14

[1]Thepigsite.com (2002) Weighing a Pig Without a Scale. https://www.thepigsite.com/articles/weighing-a-pig-without-a-scale. Accessed 05 August 2021.

CHAPTER 15

[1]Defra (2020) Code of practice for the welfare of pigs. https://assets.publishing.service.gov.uk/government/uploads/system/uploads/attachment_data/file/908108/code-practice-welfare-pigs.pdf. Accessed 28 July 2021.

[2]Asseo L (2020) Tusk trims in miniature pigs. LafeberVet website. Available at https://lafeber.com/vet/tusk-trims-in-miniature-pigs/. Accessed 10 Sept 2022.

[3]Grisha Stewart (2022) Cooperative care streaming video series. https://store.grishastewart.com/products/cooperative-care. Accessed 07 Sept 2022.

[4]The Mutilations (Permitted Procedures) (England) Regulations 2007. https://www.legislation.gov.uk/ukdsi/2007/9780110757797. Accessed 10 Sept 2022.

[5]Animal Welfare Act (2006) https://www.legislation.gov.uk/ukpga/2006/45/contents. Accessed 10 Sept 2022.

CHAPTER 16

[1]Defra and APHA (2012) Fallen stock and safe disposal of dead animals. https://www.gov.uk/guidance/fallen-stock. Accessed 09 Sept 2022.

[2]Pig Veterinary Society (2013) The casualtypPig. https://www.pigvetsoc.org.uk/files/document/192/Casualty%20Pig%20-%20April%202013-1.pdf. Accessed 09 Sept 2022.

[3]Grist A, Lines JA, Knowles TG, Mason CW, Wotton Stephen B (2018) The use of a non-penetrating captive bolt for the euthanasia of neonate Piglets. Animals. 8, 48.

[4]HSA (2016) Captive bolt stunning of livestock. https://www.hsa.org.uk/downloads/publications/captive-bolt-stunning-of-livestock-updated-logo-2016.pdf. Accessed 09 Sept 2022.

[5]Council Regulation (EC) No. 1099/2009 on the protection of animals at the time of killing. https://www.legislation.gov.uk/eur/2009/1099/contents. Accessed 05 Nov 2022.

[6]Regulation (EC) No 853/2004 of the European Parliament and of the Council. https://www.legislation.gov.uk/eur/2004/853/contents. Accessed 05 Nov 2022.

[7]Animal Welfare Act (2006) https://www.legislation.gov.uk/ukpga/2006/45/contents. Accessed 10 Sept 2022.

[8]HSA (2018) On-farm slaughter of livestock for consumption. https://www.hsa.org.uk/downloads/technical-notes/tn8-on-farm-slaughter-of-livestock-for-consumption.pdf. Accessed 09 Sept 2022.

GLOSSARY

[1]EU 2017/302. https://eur-lex.europa.eu/legal-content/EN/TXT/PDF/?uri=CELEX:32017D0302&from=EN. Accessed 20 August 2022.

ABOUT THE AUTHORS

Claire Scott BVetMed MRCVS
Claire developed her love for smallholder pigs whilst practising as a pig vet in the South West of England. She has delivered pig teaching across UK veterinary schools, to vets in practice, and to both producers and pet pig keepers. She is now completing her PhD at the University of Bristol, exploring how smallholders prevent, diagnose and treat disease.

Claire with Sue and Stephen's boar Bran.

James Adams BVSc DipACVIM (LAIM) PG Cert Vet Ed (FHEA) MRCVS
James began his career in private farm-animal veterinary practice, after which he has spent the past ten years teaching at academic institutions where he aims to ensure that vets of the future receive dedicated teaching for pet pigs. James is a diploma holder in Large Animal Internal Medicine from the American College of Internal Veterinary Medicine.

James taking care of one of his many piggy clients.

Peter Siviter BVetMed MRCVS

Pete is a farm vet from Dorset. A neighbour's orchard pigs first sparked his interest in this species, and he has spent the last ten years at Synergy Farm Health working closely with small-scale pig producers, pedigree breeders and pet pig owners. Amongst his usual veterinary work Pete stewards pigs at the county show, writes countless articles, and delivers pig training for other vets.

Pete stewarding the pig classes at the Dorset County Show.

Sue and Stephen Dudley

Stephen and Sue own Black Orchard Large Blacks, a successful enterprise producing very high welfare, soya-free, rare-breed pork, and selling breeding and meat weaners. They keep the rarest lines to help preserve the wonderful Large Black breed. Passionate about pig welfare amongst smallholders, they also run courses for new pig keepers. You can find out more at www.black-orchard.co.uk.

Sue and Stephen with Blodeuwedd, their lovely and rare Bess sow, one of only six in her line.

INDEX

African Swine Fever (ASF) 28, 95-96
all-in-all-out (AIAO) system 32
anthrax 98
antibiotics 130-131
antiparasitic medicines 131
  use of 144-146
Ascaris suum, life cycle of 114-115
assurance schemes
  labelling 24-25
  organic assurance scheme 23
  Red Tractor certification 23
  RSPCA Freedom Foods 23-24
  Wholesome Food Association 24
Aujeszky's disease 97
avermectins 144

bacterial diseases 94
bedding 39
benzimidazoles 144
Berkshire pig 15-16
biosecurity
  APHA-approved isolation facility 35
  disinfectants, use of 33-34
  for showing pigs 34-35
  on to a holding
    all-in all-out (AIAO) system 32
    isolation area 30
  incoming people and objects, holding 32-33
blocked urethra 124
boar taint 11
body-condition score 52-53
Brachyspira hyodysenteriae 107
bracken toxicity 101
breeding
  birth notification 179
  farrowing
    immediate post-farrowing period 166-173
    internal examination 164
    oxytocin release 160
    oxytocin use 165
    post-farrowing to weaning 175-178
    problems 163-166
    not giving birth 163
    obstruction 164
    stressed sows 164
    uterine exhaustion 164
    Prostaglandin F2a 159
    stages of 159-163
  hand-rearing orphan piglets 173-175
  pregnancy
    problems 157-158
    scanning 154-155
    visual signs of 155-157
  registration of pigs 179-180
  selecting for
    bought-in gilts 150
    home-bred gilts 148-150
  service
    artificial insemination (AI) 152-154
    natural service 152, 153
  signs of heat 151
  sourcing breeding stock 147-148
  sow's heat cycle 150-152
  weaning 178
Breed Societies 26
British Landrace pig 16
British Lop pig 16-17
British Pig Association (BPA) 26

cancers 124
Classical Swine Fever (CSF) 96
coccidiosis, controlling 146
Code of Practice for the Welfare of Pigs 8
colostrum 169-173
common diseases 98
creep feeding 56

death and euthanasia
  firearms and shotguns 206-207
  larger pigs
    by lethal injection 204
    penetrative captive bolt stunner 205-206
  methods used for euthanasia 203
  non-emergency cases 207
  piglets
    by blunt force trauma 203-204
    by lethal injection 203
    non-penetrative captive bolt stunner 204
diseases
  bacterial diseases 94

blocked urethra 124
cancers 124
external parasites
lice 119-120
scabies 119
fungal diseases 94
gastrointestinal disease
atresia ani 109
causes 105
Clostridium perfringens 106
coccidia 106-107
E. coli 105
hernias 110
ileitis 108
intestinal parasitic worms 108-109
intestinal torsion 109
Post-Weaning Multisystemic Wasting Disease (PMWS) 108
rotavirus 105-106
salmonellosis 107
stomach ulcers 109-110
supportive care 105
swine dysentery 107-108
Trichuris suis (whipworm)
signs of 109
control of 144

generalised disease
bracken toxicity 101, 102
erysipelas 98-99
vaccination 139-140
Glässer's disease 99-100
heat stroke 101
iron deficiency anaemia 100
Mulberry heart disease 100
mycotoxicity 100-101
rodenticide toxicity 101
vitamin E and selenium deficiency 100
Greasy Pig disease 118
infectious diseases 93-94
Kunekune spring moult 120
lameness and leg issues
foot problems 113
fractures 112-113
joint ill 111-112
joint infection in older pigs 112
non-infectious arthritis 112
liver disease
Clostridium novyi 117-118
coal tar poisoning 118
neurological disease
E. coli bowel oedema 103
Haemophilus parasuis 103
hemlock toxicity 104
lead toxicity 104
meningitis 102
nightshade toxicity 104-105
Streptococcus suis 102-103
water deprivation/ salt poisoning 104
non-infectious diseases 93
notifiable disease
African Swine Fever (ASF) 95-96
anthrax 98
Aujeszky's disease 97
Classical Swine Fever (CSF) 96
Foot and Mouth Disease (FMD) 96-97
Porcine Epidemic Diarrhoea Virus (PEDv) 97
Swine Vesicular Disease (SVD) 97
Teschen disease 97
Tuberculosis (TB) 98
parasites 94
Porcine Dermatitis Nephropathy Syndrome (PDNS) 118-119
prevention 30
of reproduction
acute mastitis 123
leptospirosis 122
metritis 123-124
parvovirus 121
Porcine Respiratory and Reproductive Syndrome Virus (PRRS) 121
Postpartum Dysgalactia Syndrome (PDS) 122-123
prolapse 122
reproductive loss 122
urinary tract infections 124
respiratory diseases
Actinobacillus Pleuroneumoniae (APP) 116-117
Ascaris suum 114
atrophic rhinitis 117
Enzootic pneumonia (EP) 116
Porcine Circovirus Type 2 116
Porcine Respiratory and Reproductive Syndrome Virus (PRRS) 114-116
secondary bacterial pneumonias 117
swine influenza 116
sunburn 120
viruses 94
zoonotic diseases 94-95
disinfectants, use of 33-34
Durocs 17-18

eAML2 system 27, 73
ear notching 194-196
ear tagging 189

E. coli
gastrointestinal disease 105
neurological disease 103
vaccines 141
electronic medicines book (eMB) 28
enzootic pneumonia (EP) 116
external parasites
lice 119-120
prevention of 143
scabies 119

Faecal Egg Counting (FEC) 144-145
farrowing
accommodation 45-47

good hygiene at 111
immediate post-farrowing period 166-173
post-farrowing to weaning 175-178
problems 57, 163-166
sows infected with lice/mites before 146
stages of 159-163
vaginal discharge 163
feeding
balance of diet 52-54
choice of diet
commercial diets 54-55
supplementation 55
for growth
post weaning 56-57
pre weaning 55-56
key ingredients
fats, oils and carbohydrates 50
fatty acids 50
fibre or roughage 51
protein 50
vitamins and minerals 50-51
water 50
legislation 28
mycotoxins 51
pet pigs 59-61
provision of food 38
sows for reproduction 57-59
fencing 47-49
Five Freedoms framework 8
flushing 59, 152
Foot and Mouth Disease virus (FMD) 96-97
foot trimming 198-199
fungal diseases 94

gastrointestinal disease
atresia ani 109
Clostridium perfringens 106
Coccidia 106-107
E. coli 105
hernias 110
ileitis 108
intestinal parasitic worms 108-109
intestinal torsion 109
Post-Weaning Multisystemic Wasting Disease (PMWS) 108
rotavirus 105-106
salmonellosis 107
stomach ulcers 109-110
swine dysentery 107-108
generalised disease
bracken toxicity 101, 102
erysipelas 98-99
Glässer's disease 99-100
heat stroke 101
iron deficiency anaemia 100
mycotoxicity 100-101
rodenticide toxicity 101
vitamin E and selenium deficiency 100
Gloucester Old Spot pig 18
Greasy Pig disease 118

handling pigs
forward planning 64-65
patience 65
positive interactions 64
restraining pigs 66-67
bucket of feed 67
by holding 67
sedation or general anaesthetic 72
snares 69-71
hog cholera *see* Classical Swine Fever (CSF)
hormones 131

infectious diseases 93-94
signs of 35
injections 132-137
internal parasites 142
isolation facility 35

Kunekunes 18-19

labelling of pork and pork products 25
lameness and leg issues
foot problems 113
fractures 112-113
joint ill 111-112
joint infection in older pigs 112
non-infectious arthritis 112
Large Black pig 19
Large White pig 19-20
legislation and regulation
feeding 28
herd registration 27
land management 29
land registration 27
medicines 28
movements and transport 27
pig identification 28
selling produce 29
sick and deceased pigs 29
standstill periods 28
walking 29
welfare 28-29
liver disease
Clostridium novyi 117-118
coal tar poisoning 118

Mangalitza pig 20-21
medicines
accidental self-injection 137
administering injections 132-137
administering oral medicines 131-132
antibiotics 130-131
antiparasitic medicines 131
broken needles 137
hormones 131
estimating weight 133
injections 132-137
injection aids 133-134
needle sizes 135
site of injections 135

pain and inflammation
non-steroidal anti-inflammatory drugs (NSAIDs) 129-130
steroids 130
rules and regulations around medicines
buying and disposing 128
licensing 126
medicines records 128
prescription-only medicines (POM) 125-126
storing medicines 127-128
withdrawal periods 126
Menter Moch Cymru 26
Middle White pig 21
movement licenses 73

neurological disease
E. coli bowel oedema 103
Haemophilus parasuis 103
hemlock toxicity 104
lead toxicity 104
meningitis 102
nightshade toxicity 104-105
Streptococcus suis 102-103
water deprivation/ salt poisoning 104
non-infectious diseases 93
non-steroidal anti-inflammatory drugs (NSAIDs) 129-130
nose ringing 200

oral medicines 131-132
organic pig production 23
Oxford Sandy and Black pig 21-22

parasite control
coccidiosis 146
external parasites, prevention of 143
hygiene and land management 142
internal parasites 142
therapeutics
avermectins 144
benzimidazoles 144
Faecal Egg Counting (FEC) 144-145
use of antiparasitic medicines 144-146
parasites 94
pet pigs
animal sanctuary checks 13
feeding 59
housing 49
sourcing 12
suitable environments 12
pig choices
assurance schemes
labelling 24-25
organic assurance scheme 23
Red Tractor Certified Standards 23
RSPCA Freedom Foods 23-24
the Wholesome Food Association 24
breeder choice
meat characteristics 15
rare breeds 14-15
rooting 14
size and character 14
small-scale producer 10-12
society memberships
breed societies 26
British Pig Association (BPA) 26
smallholding vet schemes 25
social media groups 25
in Wales 26
sourcing pigs 8-10
the Berkshire 15-16
the British Landrace 16
the British Lop 16-17
the British Saddleback 17
the Durocs 17-18
the Gloucester Old Spot 18
the Kunekune 18-19
the Large Black 19
the Large White 19-20
the Mangalitza 20-21
the Middle White 21
the Oxford Sandy and Black 21-22
the Tamworth 22
the Welsh 22-23
pig housing and environment
electric fencing 47-48
environmental fundamentals
bedding 39-40
biosecurity 44
drinking water 37-38
food 38-39
natural behaviours, expression of 45
protection from weather 41-42
shelters 40-41
farrowing accommodation 45-47
house pig 49
indoor and outdoor systems 36-37
space requirement 47
stock fencing 47
pig to pork
to the abattoir
to and from 185-186
collection 187
cutting lists 186-187
loading and travel 184-185
movement order 184
storage 187-188
on arrival 182
collection of meat 181-182
preparations before collection 181
plum pudding pig 21
Porc Blasus 26
Porcine Circovirus Type 2 116
Porcine Dermatitis Nephropathy Syndrome (PDNS) 108, 118-119
Porcine Epidemic Diarrhoea Virus (PEDv) 97
Porcine Respiratory and Reproductive Syndrome Virus (PRRS) 114-116
Post-Weaning Multisystemic Wasting Disease (PMWS) 108
procedures
detusking 196-198
ear notching 194-196
ear tagging 189

foot trimming 198-199
identification of pigs 189
neutering
castration 200-201
spaying 201
nose ring 200
slap marking 190-192
tattooing 192-194

Red Tractor Certified Standards 23
respiratory diseases
Actinobacillus Pleuroneumoniae (APP) 116-117
Ascaris suum 114
atrophic rhinitis 117
Enzootic pneumonia (EP) 116
Porcine Circovirus Type 2 116
Porcine Respiratory and Reproductive Syndrome Virus (PRRS) 114-116
secondary bacterial pneumonias 117
swine influenza 116
restraining pigs 66-72
rodenticide toxicity 101
RSPCA Freedom Foods 23-24

Saddleback 17
slap mark 190-192
snaring a pig 68-71
sourcing pigs 8-10
FAWC's essentials of stockmanship 10
the Five Freedoms 8
standstill periods 28
steroids 130
stocking densities 47
strongyle worms 109
sunburn 120
swine dysentery 31, 107-108
swine influenza 116
Swine Vesicular Disease (SVD) 97

Tamworth 22
tattooing 192-194
Teschen disease 97
transport of pigs
administrative legislative requirements
animal transport certificate 74
movement licences 73
standstill period and identification 74
fitness for transport 77-78
practical tips for transporting pigs 74-77
Tuberculosis (TB) 98

vaccines
clostridial bacterial disease 141
E. coli 141
Enzootic Pneumonia (EP) 141
erysipelas 139-140
infectious diseases 141
parvovirus 140-141
PCV2 virus 141
Porcine Reproductive and Respiratory Syndrome (PRRS) 141
storage of 139
vaccination schedule 140
vet, working with your
aftercare 88-89
behavioural changes 81
cardiovascular and respiratory systems, changes in 81-82
clinical signs, to contact 84-85
defecation and urination 82
first aid 87
identifying disease in pigs
locomotion changes 82
pain and pain management in pigs 91-92
record keeping 80
skin changes 83
streamlining the vet visit 86-87
supportive care 87-88
temperature 83
veterinary health plan 90-91
weight loss 82
viruses 94

walking pigs (legislation) 29
wallows 42-43
water consumption and provision 37-38
Welsh pig 22-23
The Wholesome Food Association 24
withdrawal period 126

zinc 51
zoonotic diseases 94-95